STARS AND PLANETS
of the Southern Hemisphere

What the critics have said about this book:

"... an outstanding introduction to astronomy and practical observing." *Christchurch Press*

"... well produced and extremely well written ... a credit to yourself and to New Zealand astronomy in general." *Carter Observatory, Wellington*

"Is it my imagination or are books being packaged more attractively these days? The cover is quite striking .... The book is well illustrated with pictures and diagrams, some in colour, and the chapters cover a wide field." *Listener*

Front cover: Voyager 1 approaches Saturn. The rings are seen nearly edge-on, with cloud belts faintly visible through the upper atmosphere haze. – *NASA photograph.*

Back cover (top): The surface of Mars from Viking 2. This rock-strewn "Marsscape" is thought to be either the eroded surface of an ancient lava flow or a flat surface littered with debris from a nearby impact crater. – *NASA photograph. (Bottom)*: The Great Nebula in Orion. This cloud of glowing hydrogen is found in Orion's Sword and is a birthplace of stars. – *U.K. Schmidt Telescope Unit, Sydney.*

STARS AND PLANETS
of the Southern Hemisphere

LIONEL WARNER
M.Sc., F.R.A.S.

ANGUS & ROBERTSON PUBLISHERS
London • Sydney • Melbourne

To
Vivienne
my wife and helpmate

Published in Australia by Angus & Robertson Publishers, Australia, 1983.
First published in New Zealand by A. H. & A. W. Reed 1978.
Reprinted 1979, 1983.

ISBN 0207 145504

Typesetting by Printset Processes (1973) Ltd, Christchurch
Printed by Kyodo-Shing Loong Printing Industries Pty Ltd, Singapore

CONTENTS

ACKNOWLEDGMENTS

As I take up my pen to acknowledge the encouragement received while writing this book, I think of many persons and organisations.

I also give thought to the astronomically minded friends whom I have never met but who let me know by letter or telephone that they read or listen to my books, newspaper articles or radio programmes.

My gratitude goes to them all, even though limitations of space deny me the opportunity of mentioning more than just a few of their names.

However, included among those whom I thank particularly are: Mr Frank Clift, former President of the Royal Astronomical Society of New Zealand; Mr Ian Hayter, Principal of Lynfield College; Professor Les Allen of the University of Hawaii; my sister Sylvia Ashton-Warner; Mr E. S. Curtis; Mr G. L. Loftus; Mr Vernon Sale and Mrs June Sale; Mrs Jackie St George; Mrs Trixie Stewart; Mrs Rosa Tyson; the President and Council of the Auckland Astronomical Society.

For supplying photographs I thank Mr A. L. Rounthwaite (many of his photographs are in this volume); Mr R. North; Mr A. B. Raphael; Mr W. Thrush; Mr B. Ward; Mr H. O. Williams; Carter Observatory; Hale Observatories; High Altitude Observatory; Lick Observatory; Lowell Observatory; Mount Stromlo and Siding Spring Observatory; California Institute of Technology; National Aeronautics and Space Administration; Air New Zealand; the newspapers New Zealand Woman's Weekly, Auckland Star, Taranaki Daily News.

Mr Julius Petro, in the Production Department of the publishers A. H. & A. W. Reed Ltd., has earned my admiration for his art work appearing in this book.

I sincerely thank my son Dr Leigh A. Warner for preparing most of the drawings and star charts, and my wife Vivienne for her unfailing encouragement.

Lionel Warner

Auckland
New Zealand

INTRODUCTION

With this book as a friendly guide, the newcomer to astronomy looks overhead to comprehend the beauty and grandeur of the heavens.

During his first exciting evening as a stargazer he makes a visual tour of the southern sky, viewing the famous Southern Cross and Pointers, the Diamond Cross, the False Cross, the space-navigational star Canopus, bright Achernar and solitary Peacock. He also discovers our two nearest galaxies, the mighty Magellanic Clouds.

On subsequent nights he locates the principal stars of the northern sky. In due course he examines the intriguing planets Mercury, Venus, Mars, Jupiter and Saturn. Frequently a meteor darts across his field of vision. On some occasions an awesome comet makes its appearance.

This book describes and explains many fascinating celestial objects. It gives individuality to the stars and planets which have earned a place in legend and history. Their nature has been revealed by earth-bound observations and by detailed reports from manned and unmanned spacecraft. The final chapter briefly takes the reader beyond the solar system into regions occupied by mysterious white dwarf stars, pulsars, black holes and quasars.

The Author. Lionel Warner M.Sc., F.R.A.S., a former head of the science department in a large secondary school, has the rare gift of explaining astronomy in simple language while conveying the spirit of his own enthusiasm for the night sky.

This ability has been developed through many years of presenting the subject in his books, newspaper articles, radio and television programmes. His major book, **Astronomy For The Southern Hemisphere,** was referred to by the internationally known magazine **Sky And Telescope** as "... this fine work". In 1976 the Royal Astronomical Society of New Zealand awarded to Lionel Warner the Murray Geddes Memorial Prize for his "long and meritorious service in the field of astronomical education".

Plate 1. The Pleiades star cluster, often called the Seven Sisters.
Carter Observatory

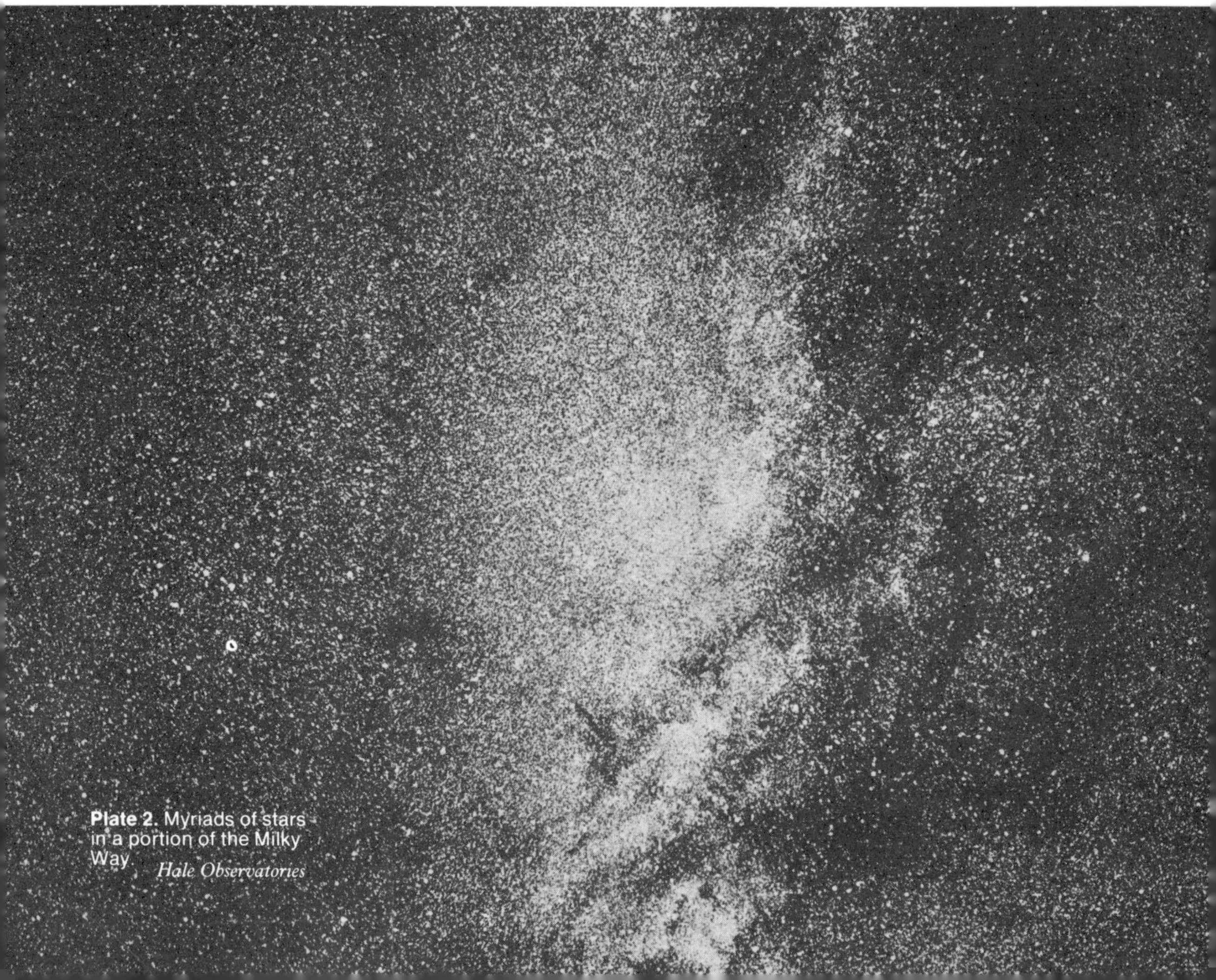

Plate 2. Myriads of stars in a portion of the Milky Way.
Hale Observatories

The Space Shuttle orbiter, called **Enterprise**, separates from a Boeing 747 aircraft which had carried it "piggyback" to a launching altitude of 8100 metres. In due course, orbiters of this type will not be assisted by a Boeing 747, but they will take off with their own rockets, perform as spacecraft in orbit round the earth and then glide down to an ordinary runway on earth.

Enterprise opens a new era in which increasing numbers of human beings will experience the excitement of journeys into space.

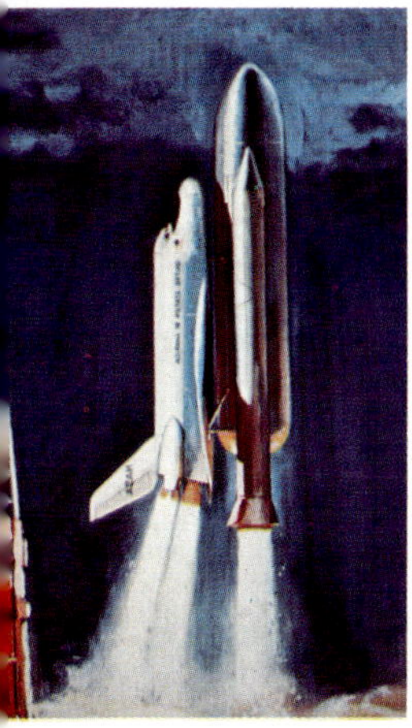

Rocket assisted lift off to a height of 30-40 km

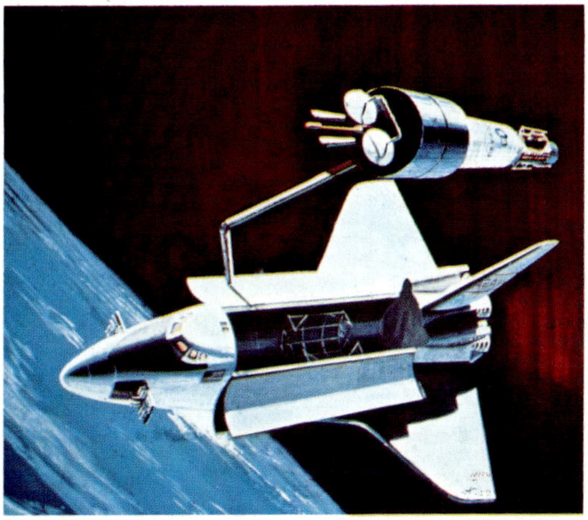

The space shuttle at work, retrieving a satellite.

Approach for a conventional landing, accompanied by a chase aircraft.

All photos by: USIS

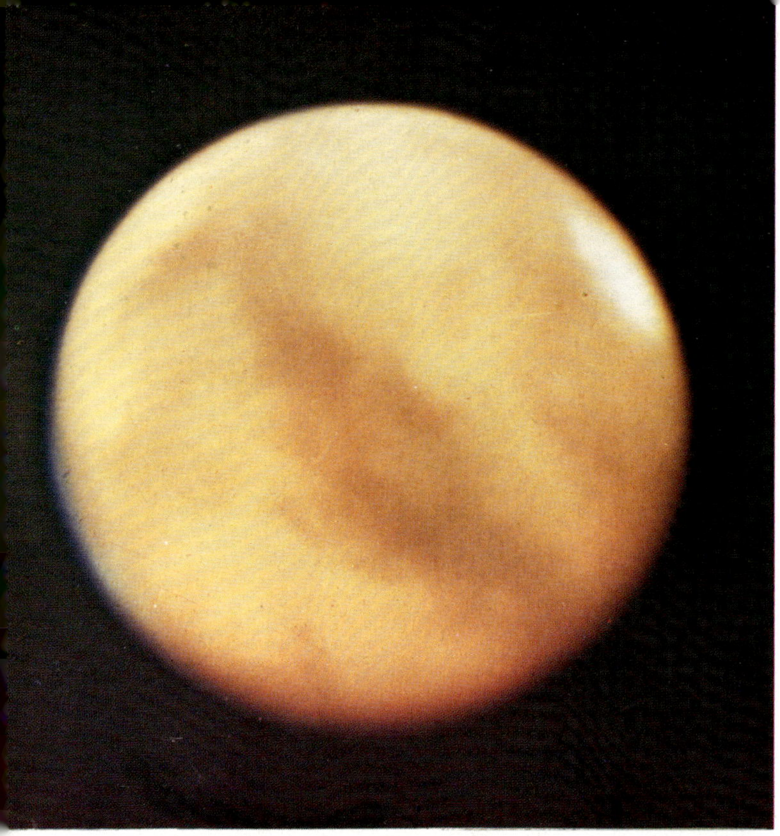

The "red" planet Mars. This telescopic view shows patches which were named in the mistaken belief that they were "seas" (dark) and "continents" (light coloured). One polar icecap is visible. Volume of Mars: one-seventh that of earth. Mean distance from the sun: 227.94 million km.

Close-up view of a Martian landscape photographed from the Viking 2 lander resting on the planet's surface. The rocks, some as large as one metre across, are porous and sponge-like, similar to certain volcanic rocks on earth.

The surface is blanketed with fine-grained material which has been spread by Martian winds. *Photo: NASA*

The giant planet Jupiter. The parallel zones of colour are cloud formations providing evidence of Jupiter's rotation of less than 10 hours each turn. The mysterious Great Red Spot seen in the upper left of the picture has a maximum diameter of 40,000 km which is more than a quarter of the diameter of the planet itself.
Volume of Jupiter: 1,318.7 times that of earth. Mean distance from the sun: 778.34 million km.

The ringed planet Saturn. The beautiful ring system is about 68,000 km from the inner circumference to the outer circumference but it is very thin (not more than a few tens of km).
Volume of the planet (not including the rings): 743.6 times that of earth. Mean distance from the sun: 1427.1 million km.

The Lagoon Nebula

I. INTRODUCING STARS

Chapter 1.

THE NIGHT SCENE

Most of us experience a sense of awe when gazing up at the night sky with its stars extending in all directions towards infinity.

In that vast perspective the earth is just a speck of dust and we, as physical entities, are tinier still.

Yet in the field of the intellect man is a giant. He has acquired much knowledge of the stars and his mind is free to roam the universe.

Fortunately we do not need to be astronomical experts to make some imaginary journeys through the sky. Even a newcomer can become acquainted with the brightest and most famous stars and planets in the course of a few evenings of observation.

All that is required, in addition to a spirit of investigation, are a few simple instructions and explanations such as are presented in the following pages.

Let us start by asking –

WHAT IS A STAR?

An average star, such as the sun, is a ball of fire blazing with the violence of countless hydrogen bombs. It is big enough to hold more than a million earths. The stars other than the sun are at such enormous distances from us that they look like mere pinpoints of light.

HOW MANY STARS?

There seem to be millions of them as well as many luminous patches which could be either nebulae (cloudlike formations of gas and dust) or concentrations of stars too close together in our line of sight for recognition as individual stars.

However, careful counts have shown that the number of separate stars we can see in our half of the sky is only about 3000, though, of course, a telescope raises this number to myriads. No one knows where the universe ends or whether it ever ends at all.

The British astronomer Sir Arthur Eddington (1882-1944) wrote: "We know the prodigality of Nature. How many acorns are scattered for one that grows to an oak? And need she be more careful of her stars than of her acorns?"

HOW FAR FROM EARTH?

Stars are so remote that we cannot express their distances in kilometres without using awkwardly big figures. Even the closest of stars, not counting the sun, is 41 million million km away from us – that is the star Alpha Centauri.

To avoid the difficulty, astronomers use light-years instead of kilometres. A **light-year** is the distance travelled by light in one year, namely 9.4607 million million kilometres. On that basis we can say that Alpha Centauri is 4.3 light-years from the earth. (Fig. 6, page 4). The most distant stars at the boundaries of the known universe could be more than 10,000 million light-years away and that would mean that light from them arriving here now took 10,000 million years to make the journey. In other words we are seeing those stars not as they are now but as they were 10,000 million years ago. They might already have changed beyond recognition.

WHY SHOULD WE BE INTERESTED IN STARS?

Many people add richness to their intellectual lives by learning about their physical environment – about flowers, trees, land forms, rocks, shells, fossils or other natural objects. To such people the countryside or the seashore is a friendly place where there is always something new to discover and enjoy.

But there is another part of our environment which can be equally friendly – the sky. Each of the principal stars and planets invites attention because of its individual characteristics. It has its own degree of brightness; its colour, size, and distance from us; its path across the heavens; and in some cases its story as related in the literature of the nations of antiquity.

The celestial scene is occasionally enlivened by the unexpected sighting of a meteor, comet or man-made satellite. An eclipse of the moon or the more spectacular eclipse of the sun occurs at certain intervals. From arctic or antarctic regions the beautiful Aurora Borealis or Aurora Australis can frequently be seen brightening the heavens.

Any clear night can be depended upon to provide fascinating, or sometimes exciting, views of Nature's handiwork on the grand scale.

Professional astronomers are particularly interested in the scientific study of the many types of objects scattered through space. To those highly trained men and women the stars are distant laboratories in which sub-atomic particles are interacting under extremes of pressure and temperature unobtainable on earth. Information from such sources increases man's understanding of the atom and thereby, in the long run, affects human destiny.

II. LOOKING SOUTH

Chapter 2.

THE SOUTHERN CROSS AND ITS NEIGHBOURS

THE SOUTHERN CROSS

Many a navigator through the ages has looked up at a particular area of the sky to be reassured and guided by the most famous group of stars in the southern sky – the Southern Cross.

Countless other people have also gazed at the Cross with interest, many of them no doubt recalling that it is represented on the national flags of both Australia and New Zealand.

Any newcomer wishing to locate the Cross needs only to examine the sky in the general direction of south and, if he knows what to look for, he should soon spot it. Shortly after sunset in mid-January the Cross will be lying on its side as shown in Fig. 1. Its position in other months can be ascertained by turning Fig. 8 around until the relevant month is at the bottom of the circle (for explanation see Chapter 3).

Confirmation that the correct Cross has been found is the presence of the two bright Pointers indicated in Fig. 1.

Of course satisfactory viewing of the stars requires a clear night with little or no moonlight. The observation site should be reasonably free from the glare of city lights and there should be no appreciable visual obstruction by hills or high buildings. Astronomers favour country districts for best viewing.

Plate 3. The Southern Cross and Pointers. A convenient starting point for a visual exploration of the southern sky. *R. North*

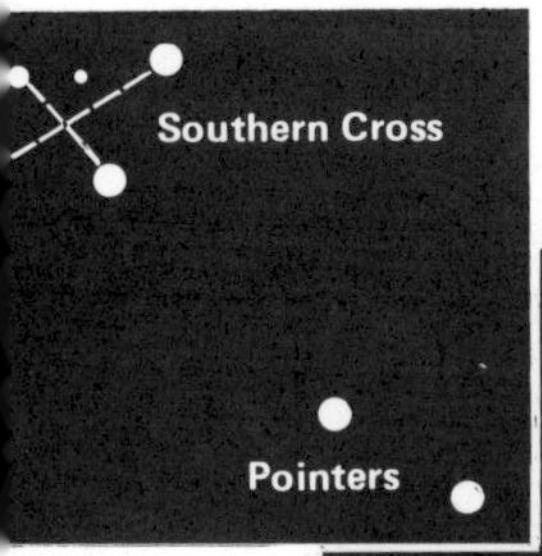

Fig. 1. The Southern Cross (5 stars) and Pointers (2 stars)

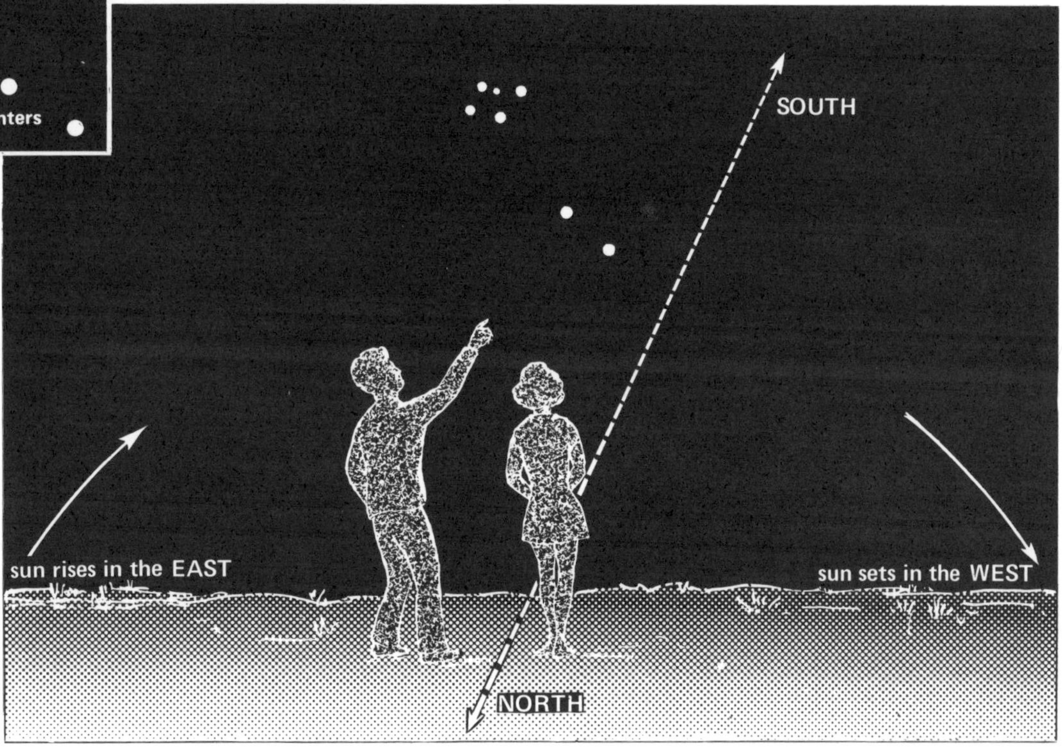

Fig. 2. To find approximate south, stand so that on your left is the position where the sun rises (in the east or not far from east), and on your right is the position where the sun sets (in the west or not far from west). Then you will be facing south, with your back towards the north. This method, though not exact, is helpful in locating the Southern Cross and Pointers for the first time.

INDIVIDUAL STARS IN THE CROSS

The five principal stars of the Cross belong to a larger group officially known as the constellation Crux, a Latin word meaning Cross. There are altogether 88 constellations in the entire sky and Crux is the smallest in area.

The brightest star in the Cross is named Alpha Crucis. Alpha is the first letter of the Greek alphabet and Crucis means of, or belonging to, Crux. The others, in descending order of brightness, are given names which include the next four letters of the Greek alphabet: Beta Crucis, Gamma Crucis, Delta Crucis and Epsilon Crucis.

Faint colours can be discerned in these stars when viewed with the naked eye or, better still, through binoculars. Gamma is red, a colour indicating that its surface is relatively cool at about 2000°C. Epsilon is orange and somewhat hotter. Alpha, Beta and Delta are bluish-white and very hot indeed with temperatures approximately 25,000°C or four times hotter than the sun.

The different brightnesses and colours of these stars give them individuality and enhance our interest in them.

Four of your fingers held together at arm's length will fit approximately into the long axis of the Cross.

Fig. 3. The stars of the Southern Cross and Pointers, with the Coal Sack.

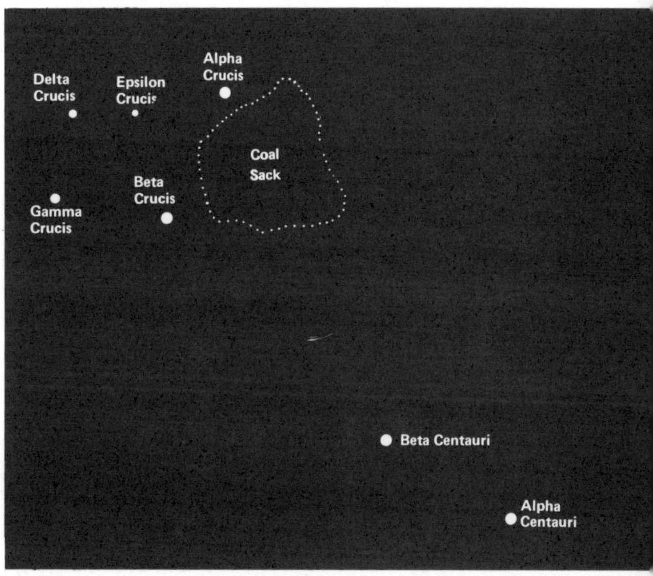

THE POINTERS

The Pointers, Alpha Centauri and Beta Centauri, are contained in the constellation Centaurus, the centaur. This mythical creature was half-horse and half-man as shown in Fig. 4. In March the constellation is conveniently placed well above the southern horizon and can be detected in the sky without much trouble.

A small telescope will reveal that Alpha Centauri consists of two stars. These components swing round their common centre of mass once every 80 years. There is also a third star which is gravitationally related to the main pair though it is relatively distant from them. At present it is closer to us than the pair and has been named Proxima Centauri, the word Proxima meaning nearest.

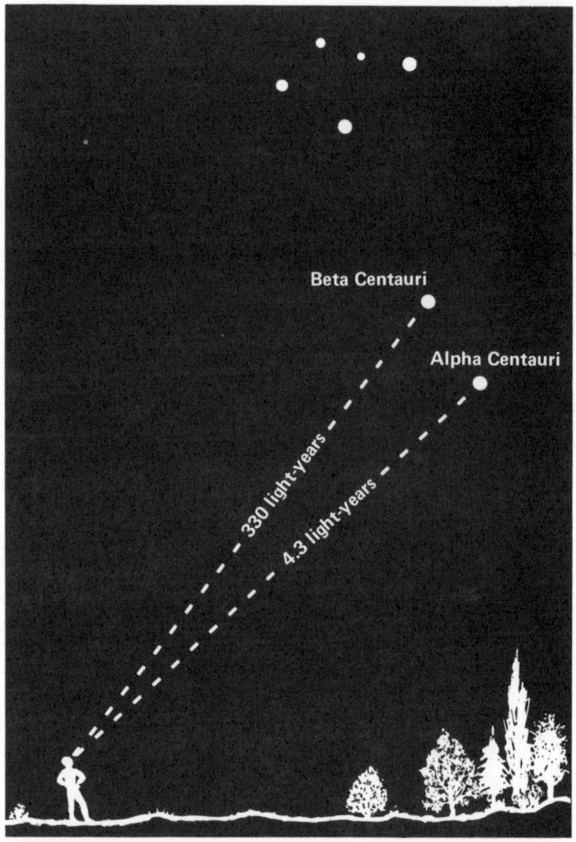

Fig. 6. Alpha Centauri is our nearest star (not counting the sun, which is also a star)

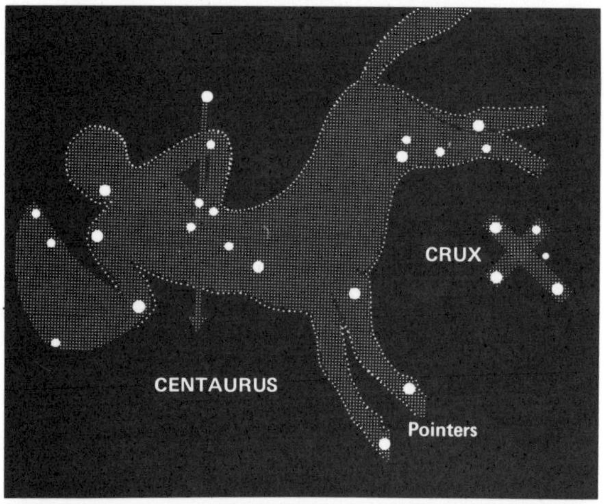

Fig. 4. The mythical centaur, half-horse and half-man, of the constellation Centaurus.

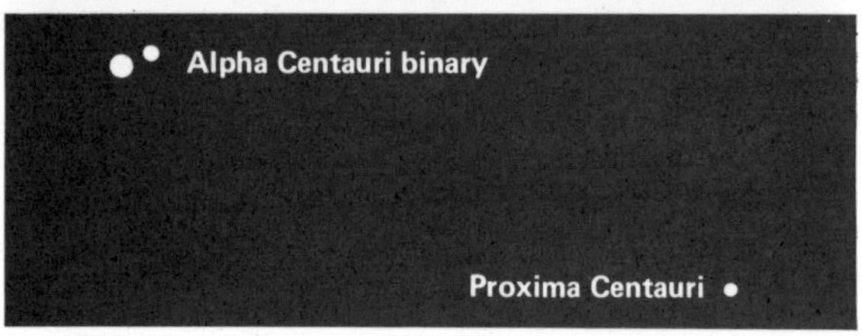

Fig. 5. The Alpha Centauri trio of stars. Their distances apart are not drawn to scale.

THREE MORE STARS, THE FALSE CROSS AND THE DIAMOND CROSS

In Fig. 8, which is a chart for January, look for the star Achernar. It can be found in the sky by taking the direction of the long axis of the Southern Cross and measuring about three handspans across the sky from Alpha Crucis.

A handspan (Fig. 7) is a rough but sometimes convenient measure of an angular distance of about 20°. To indicate a handspan, hold the fingers and thumb apart at the end of your outstretched arm. You can ascertain more accurately the number of degrees in your own handspan by counting the number of handspans measured against any background as you turn through a complete circle of 360°, and then dividing that number into 360°.

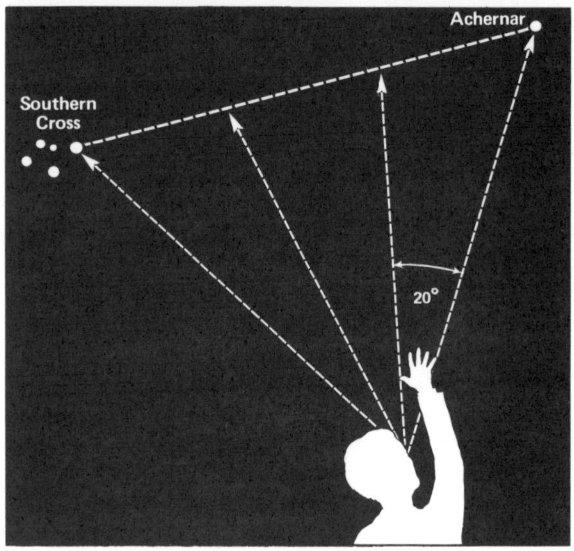

Fig. 7. The handspan is a convenient measure of approximate angular distances across the sky.

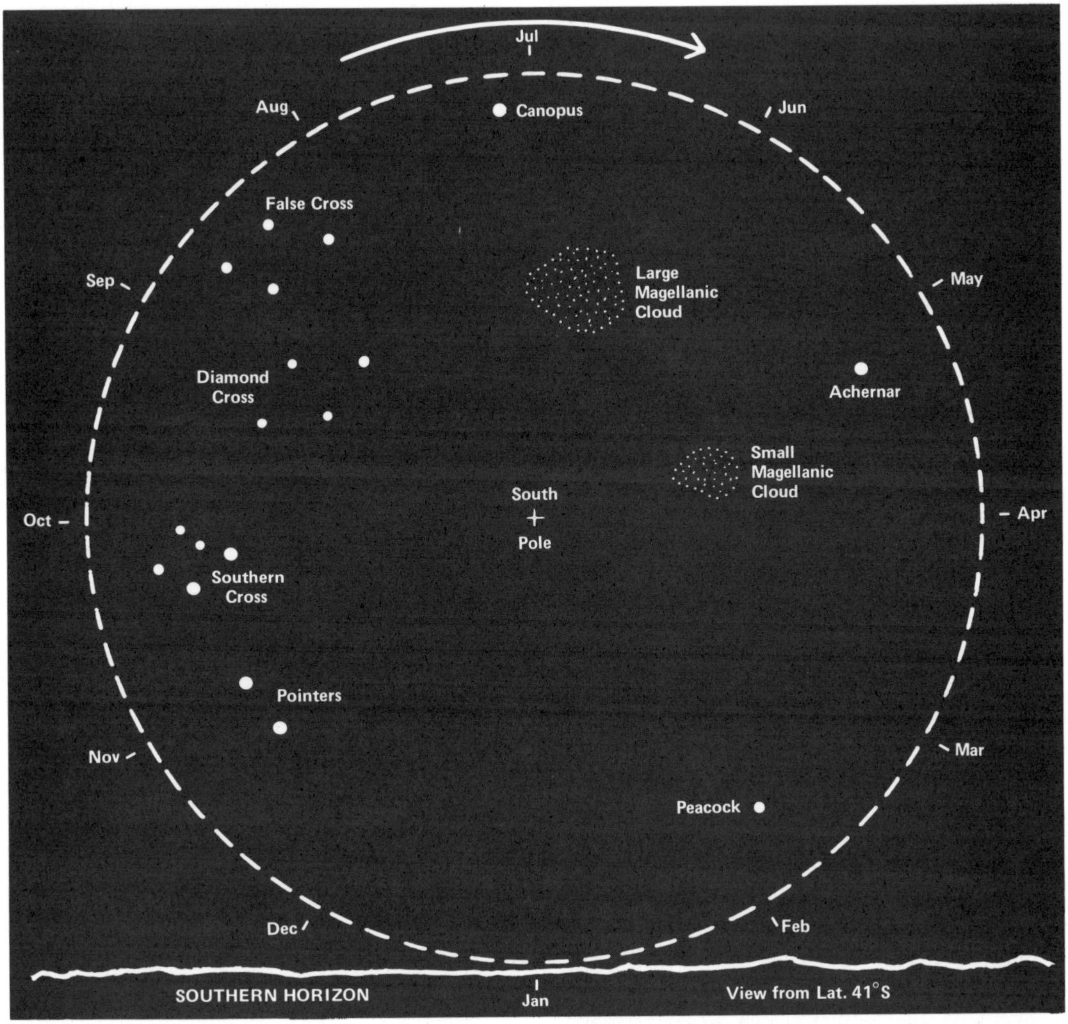

Fig. 8. The principal stars in the south polar region. As observed from about latitude 41°S, these are circumpolar stars. For any month other than January, turn the chart to bring the relevant month to the bottom of the chart.

Achernar is the ninth brightest star in the sky. Its name means End of the River, because it is situated at the end of a long "river" of stars making up the constellation Eridanus, the River. Astronomers call this star Alpha Eridani.

An imaginary straight line joining Alpha Crucis to Achernar can be regarded as the diameter of a circle indicated in Fig. 8. Looking at the figure and starting again from the Southern Cross, follow round the circle in a clockwise direction and come to the group of symmetrically placed stars known as the Diamond Cross. This is a beautiful object in the sky.

The False Cross is larger than the Southern Cross but its four stars are not so bright. When glimpsed between clouds it could be mistaken for the Southern Cross although the absence of Pointers could help to distinguish it.

Canopus, second brightest star in the sky, has always been an important guiding star for navigators and in modern times it has achieved fame in space exploration. When an unmanned spacecraft, such as those used in the Mariner and Viking projects, has left earth on a journey to nearby planets, a manoeuvre must be carried out to ensure that the craft's high-gain radio antenna points directly and continuously towards earth. This requires that the craft be stabilised in its attitude in space.

An order goes out from earth instructing the aircraft to "lock on" to Canopus. Gas jets on board set the craft rolling on a particular axis until light from Canopus enters a sensor which stops the rolling. Scientists on earth, hearing the signals suddenly coming in loud and clear, interpret them as indicating that the manoeuvre has been completed satisfactorily. Similar stabilising procedures, using other light sources, are adopted for two other axes.

Peacock is a relatively faint star but because there are no other brighter stars in its neighbourhood it is useful in star recognition.

OUR NEAREST GALAXIES

Lying on opposite sides of the line joining Alpha Crucis to Achernar are two cloud-like patches which are not ordinary clouds but galaxies – enormous groups of stars sometimes referred to as "island universes". They are the Large Magellanic Cloud and the Small Magellanic Cloud.

Although the Magellanic Clouds are about 170,000 light-years from us, they have been studied in great detail by astronomers taking long-exposure photographs with cameras attached to telescopes.

The Large Magellanic Cloud consists of something like 10,000 million stars contained within a disk-like shape resembling that of the Milky Way Galaxy depicted in Fig. 9. Its diameter is 30,000 light-years, i.e. light takes 30,000 years to traverse it. The Small Magellanic Cloud is 25,000 light-years across.

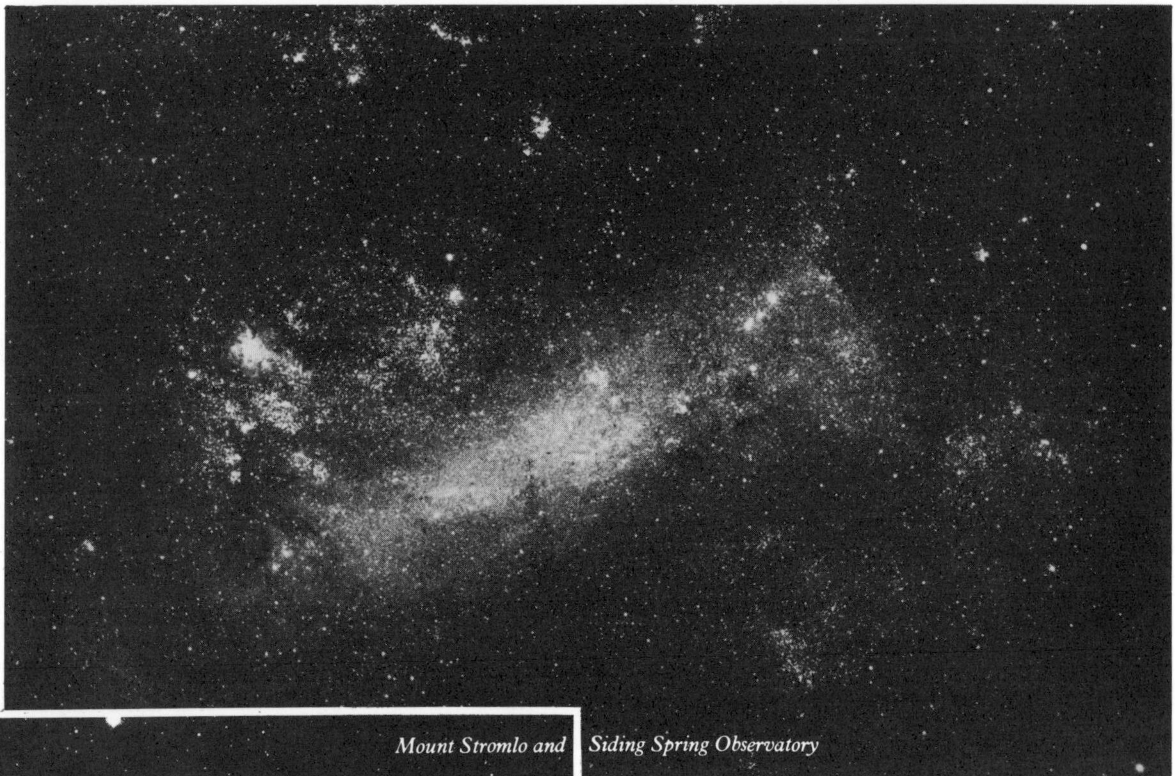

Mount Stromlo and Siding Spring Observatory

Plate 4. The Large Magellanic Cloud, composed of thousands of millions of stars. To the naked eye it appears as a wispy cloud. Light from it takes 170,000 years to reach us.

Plate 5. The Small Magellanic Cloud, a companion galaxy of the Large Magellanic Cloud.

THE BRIGHT MILKY WAY AND A DARK NEBULA

Stretching across the entire sky is an awe-inspiring band of light called the Milky Way. It is one of the great sights for stargazers and is particularly fascinating when viewed through binoculars.

The Milky Way is made of the dense conglomeration of stars we see when looking along the central axis of the mighty Milky Way Galaxy (spelt with a capital G). We ourselves, living on the earth, are within that Galaxy. Fig. 9 shows the position of the sun about two-thirds of the distance from the centre of the Galaxy to its rim. The sun's planets, including the Earth, are of course too small to be shown in the diagram.

Bigger than either of the Magellanic Clouds, the Milky Way Galaxy is 100,000 light-years across and 20,000 light-years through its central bulge.

The whole Galaxy is rotating in such a way that all of its stars are revolving at different speeds round its centre. The nearer a star is to the centre the greater its speed. Our sun is revolving at 250 kilometres per second and it takes about 225 million years for each revolution. The term **galactic year** is applied to that period of time.

Fig. 9 The Milky Way Galaxy, a vast system of stars. Within it is the sun with its family of planets including the Earth.

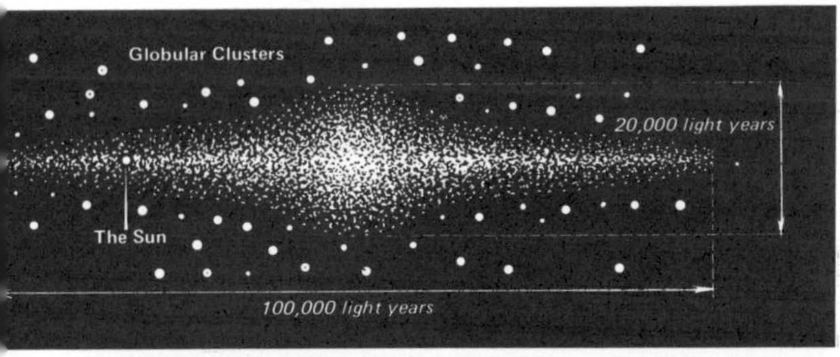

The borders of the Milky Way as we see it are irregular and the Milky Way itself is patchy in parts. Late at night when many stars throng the sky we notice that the Southern Cross seems to be embedded in the Milky Way, but this is just an effect of visibility in our line of sight. Actually the Southern Cross is much nearer to us than the background of stars comprising the Milky Way.

An object which has the appearance of a black hole in the sky is a dark nebula of which one side borders the stars Alpha Crucis and Beta Crucis. It is called the Coal Sack and is sufficiently opaque to hide the stars lying behind it.

Plate 6. This photograph of 2 hours' exposure emphasises the Southern Cross "embedded" in the Milky Way, with the dark Coal Sack bordering Alpha Crucis and Beta Crucis.

Franklin-Adams, Johannesburg

III. FOLLOWING THE STARS

Chapter 3.

HOW STARS MOVE

STAR MOVEMENTS DURING THE NIGHT

The earth is rotating under stars which are so far away that their individual movements cannot be detected except by special procedures. However, in our everyday experience the earth **seems** to be at rest and the stars **seem** to be travelling along circular paths in a direction opposite to that of the earth's rotation.

Fig. 10. The Earth turns under the stars.

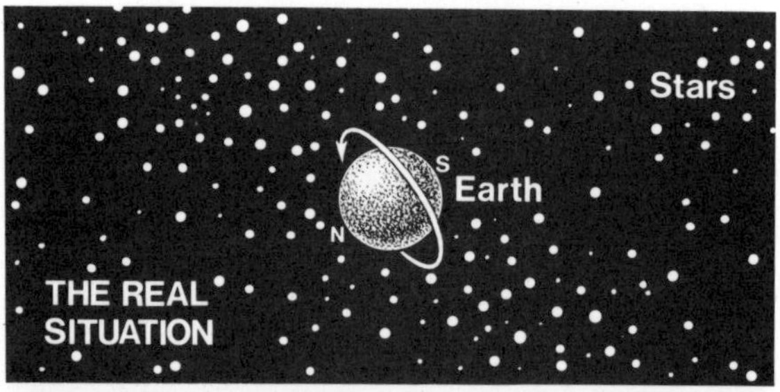

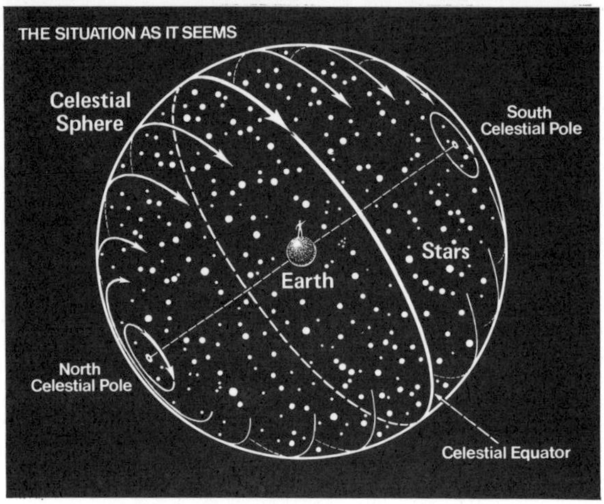

Fig. 11. Stars **appear** to revolve round a stationary Earth.

For convenience we can assume that all the stars are attached to a huge imaginary sphere called the **celestial sphere** as shown in Fig. 11 which has been drawn to represent the viewpoint of an observer in the earth's southern hemisphere. The man in the diagram is looking towards the south and he observes that the stars are travelling in a clockwise direction. If he turned round and faced north he would see the stars travelling in an anticlockwise direction.

Because the earth rotates once every 23 hours 56 minutes against the background of the stars, the stars on the celestial sphere appear to complete their circular journeys every 23 hours 56 minutes. In Fig. 11 the earth's rotational axis is extended to meet the celestial sphere at two points, the **South Celestial Pole** and the **North Celestial Pole.**

A curved arrow at the top of Fig. 8 page 5 indicates the direction in which the stars in that region are circling the South Pole. If we observe those stars in the positions shown, and then observe them again six hours later, we find that they have moved to the extent of a quarter of a circle.

STAR MOVEMENTS FROM NIGHT TO NIGHT

The stars revolve in 23 hours 56 minutes. How much further would they travel during the next four minutes to make up 24 hours?

In 23 hours 56 minutes (or 1436 minutes) the stars travel 360°. Therefore in 4 minutes they travel

$$360^\circ \times \frac{4}{1436} = \frac{360}{359} = 1^\circ \text{ approximately.}$$

Thus, if we carefully observe the position of one star, say Canopus, at 9 p.m. on one night and then carefully check its position at 9 pm on the next night, we find that it will have moved a tiny distance (actually 1°) further along its path.

Ordinarily that 1° would go unnoticed but after a month of 30 days it amounts to 30 ° and that is why, in Fig. 8, the chart has to be turned through 30° to show the star positions month by month.

After 12 months the 30° amounts to 360°, the complete circle. Hence the charts for any one month show the same star positions year after year.

CELESTIAL GLOBE

The imaginary celestial sphere is represented by an instrument called either a celestial globe or an astroglobe. It is pictured in Plate 7.

It consists of a transparent plastic sphere on the outside of which labelled "stars" have been placed in their correct relative positions. A small sphere representing the earth is supported at the centre of the transparent sphere.

The celestial globe can be used to show the positions and directions of movements of the principal stars at any time on any night of the year. It can also show the times when the sun rises and sets on any day, past, present or future.

Altogether it is a very instructive device for presenting a three-dimensional picture of the spatial relationships of the stars, the sun and the earth.

Plate 7. A celestial globe. Its transparent sphere, labelled with "stars", can be rotated to provide a three-dimensional representation of star movements.
A. L. Rounthwaite

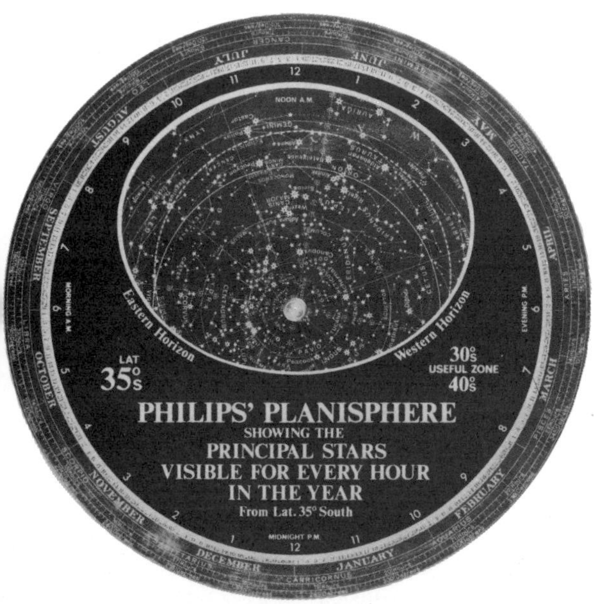

Plate 8. A planisphere, handy for indicating the positions of stars visible at any particular date and time. It applies to latitude 35°S.

PLANISPHERE

A planisphere is an inexpensive device for ascertaining what stars are visible at any time on any date. It consists of two plastic disks held, one over the other, by a central metallic dome. "Stars" on the lower disk can be viewed through a window of the upper disk which can be rotated to the required time and date printed near the boundaries of the disks.

FINDING DIRECTIONS FROM THE STARS

There could come a time in the lives of many people when, travelling at night, they require urgently to know what direction to take in order to reach their destination. This does not become a problem to trained navigators but it could concern a person adrift at sea in a small boat or a tramper lost in unfamiliar territory.

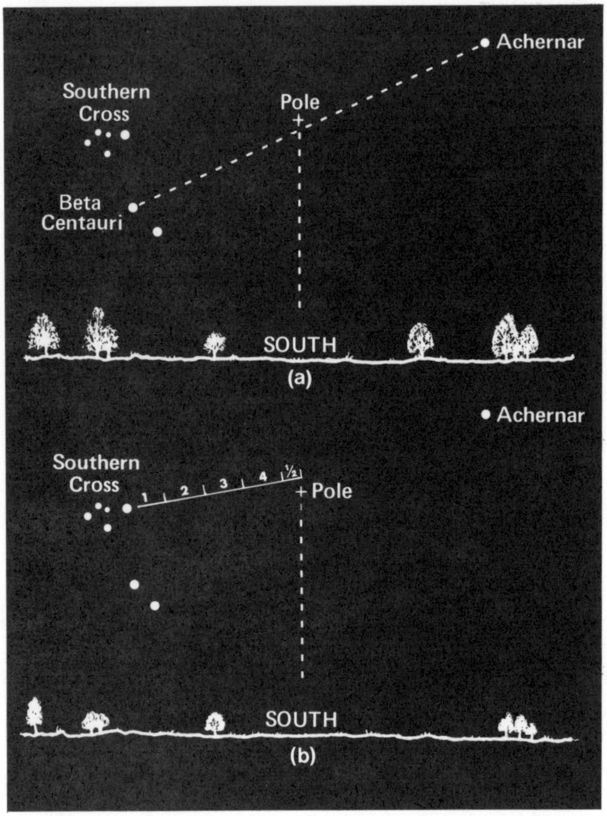

Fig. 12. Two methods of finding directions from the stars (see text).

On any clear night, certain prominent stars can serve as excellent direction-finders. Here are two examples which can be understood from a glance at Fig. 12.

1. Point one arm to Beta Centauri and the other to Achernar. Bring the arms together. They will then be aimed at a point near the South Celestial Pole. Lower the arms vertically until they are pointing to the horizon and therefore to True South.

2. Visually extend the long axis of the Southern Cross by about $4\frac{1}{2}$ times its length beyond Alpha Crucis. This will give roughly the position of the South Celestial Pole. A point on the horizon vertically below the Pole gives the direction of True South.

DIRECTIONS IN THE NORTHERN HEMISPHERE

People in the northern hemisphere have a much more straightforward method of finding a direction. They simply look towards the Pole Star (Polaris) which is almost exactly due north. It is only one degree away from the North Celestial Pole.

IV. LOOKING NORTH

Chapter 4.

SUMMER STARS

THE POT IN ORION

Clear skies in January are adorned with the constellation of Orion, famous in history and in legend.

To find it in the sky we face the general direction of north and look for a rather small group of stars which have a remarkable resemblance to a cooking pot. It is popularly known as The Pot, and it occupies the central region of the constellation shown in Plate 9. In that picture The Pot is tilted over to the left.

The bottom of The Pot is represented by three stars in almost a straight line. The handle is made up of three stars, the middle one having a fuzzy appearance. A check to identification is to note that the length of the bottom of The Pot is about half the distance between the Pointers which are near the Southern Cross.

Stargazers of the past, living in the northern hemisphere, thought that the principal stars of this constellation suggested the outline of the mythical Orion, the Hunter. We in the southern hemisphere do not get that impression unless we are prepared to accept Orion in an upside-down attitude. Fig. 15 will explain why our southern hemisphere view of star patterns in the equatorial regions of the sky is inverted compared with the northern hemisphere view.

By turning Fig. 14 upside-down we see Orion as northern hemisphere viewers see him. Astronomers call the bottom of The Pot, the Belt of Orion; the handle is the Sword of Orion.

THE STORY OF ORION

According to Greek mythology, Orion was one day hunting in the forest when a scorpion stung him on the bare heel. He fell sprawling to the ground where he soon died. But his father, the god Jupiter, brought him to life again and placed him in the sky together with his two dogs. Those dogs are represented by the constellations Canis Major (Larger Dog) and Canis Minor (Smaller Dog). In Fig. 14 Orion is seen attacking Taurus, the Bull.

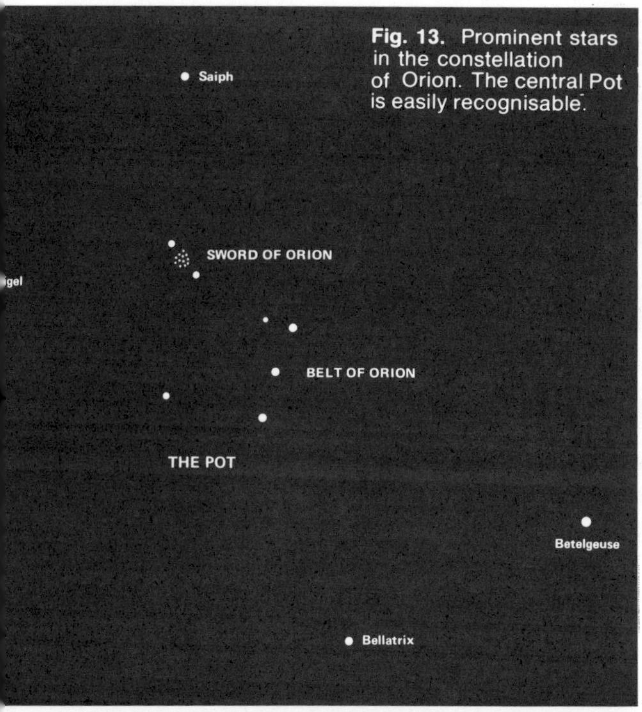

Fig. 13. Prominent stars in the constellation of Orion. The central Pot is easily recognisable.

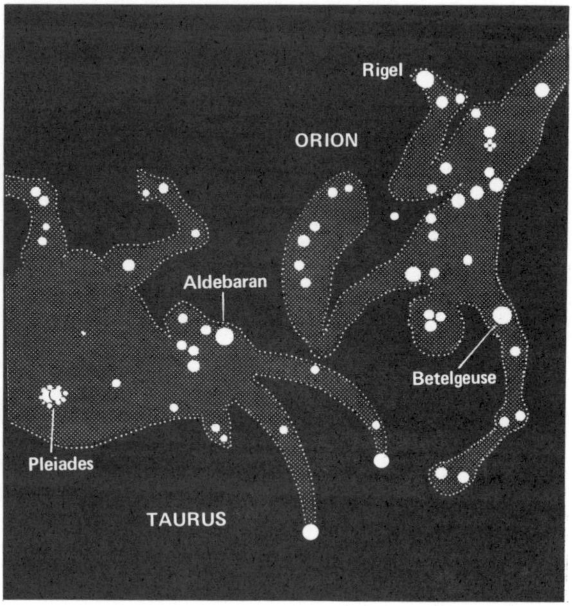

Fig. 14. Mythical figures of Orion the Hunter and Taurus the Bull in mortal combat. Compare this with Fig. 17.

Plate 9. Principal stars of the constellation of Orion.

R. North

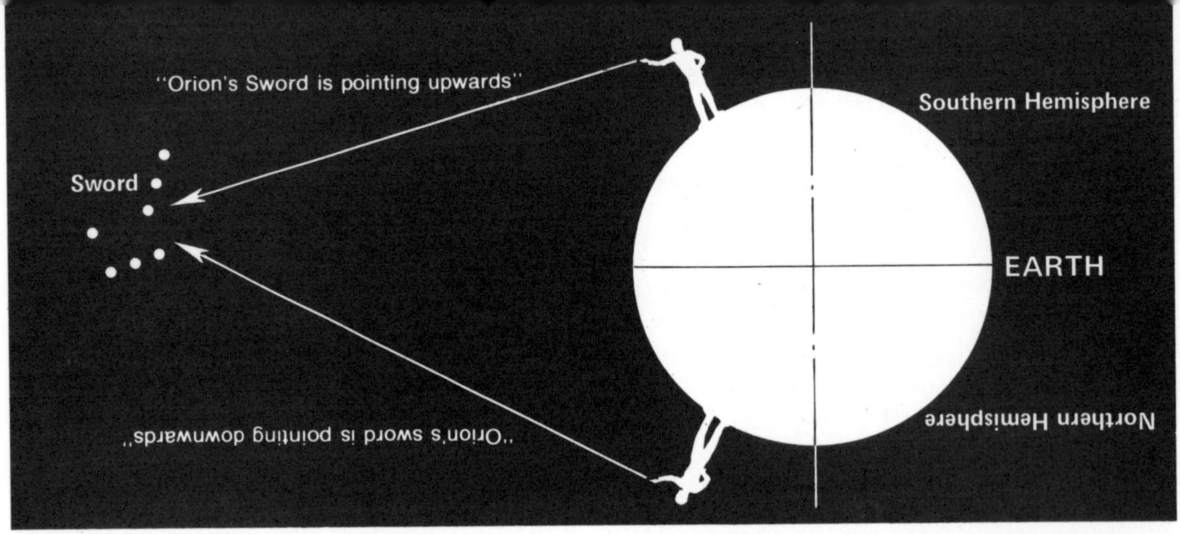

Fig. 15. Different points of view. The southern hemisphere view of star patterns is upside down compared with the northern hemisphere view, and vice versa.

MOVEMENTS OF THE NORTHERN STARS

If the man in Fig. 11, page 8 were to turn round and face north (bottom left of diagram) he would observe that the stars seemed to be travelling in an anticlockwise direction. The Orion constellation in Plate 9 has a similar apparent motion; and this applies to all the northern stars when viewed from the southern hemisphere.

FINDING DIRECTIONS FROM THE CONSTELLATION ORION

By consulting the diagrams in Fig. 16 we can understand that as The Pot travels in an anticlockwise direction it follows along close to the Celestial Equator shown as a broken circular line. (The Celestial Equator, which is a projection of the earth's Equator on the celestial sphere, meets our horizon due east and due west.)

The star Delta Orionis is almost exactly on the Celestial Equator. Therefore if a mariner sees that star beginning to rise above the horizon he knows that he is looking towards the east. If the star is just about to dip below the horizon it is due west. Also, when the handle of The Pot is upright, it is due north. A mariner experienced in star navigation could judge other directions by estimating the angle at which The Pot is tilted.

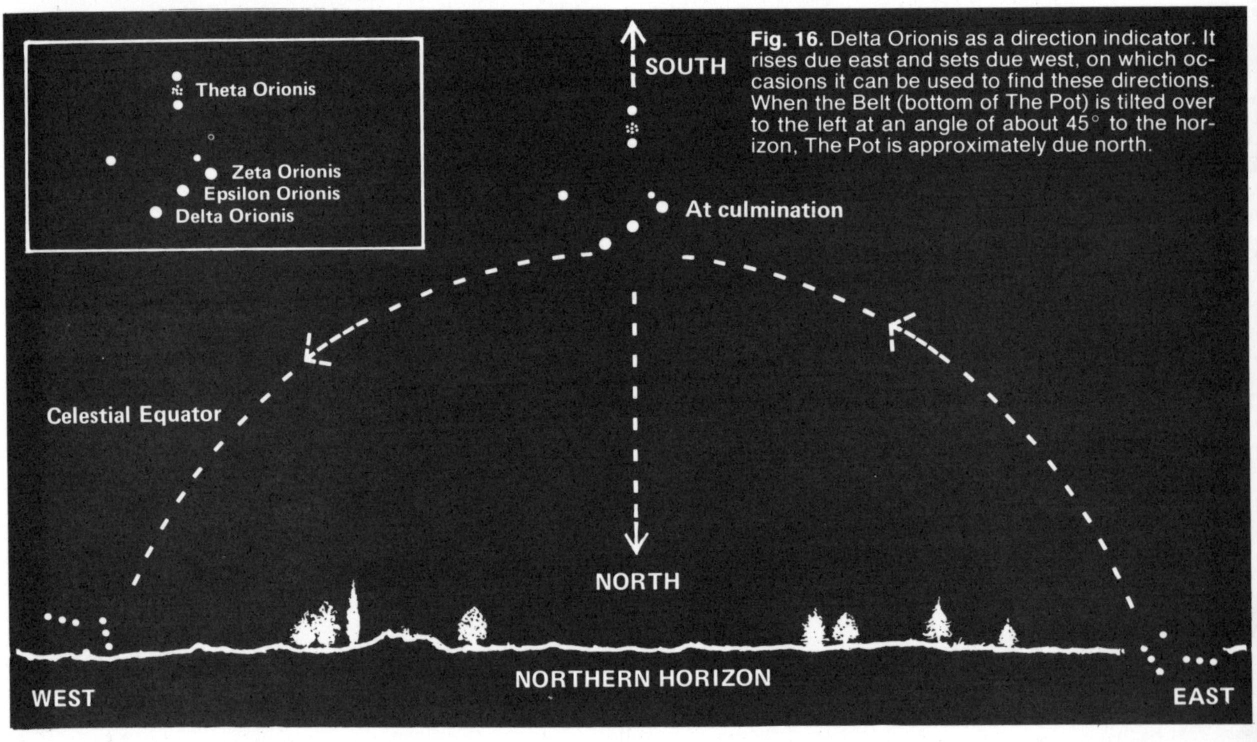

Fig. 16. Delta Orionis as a direction indicator. It rises due east and sets due west, on which occasions it can be used to find these directions. When the Belt (bottom of The Pot) is tilted over to the left at an angle of about 45° to the horizon, The Pot is approximately due north.

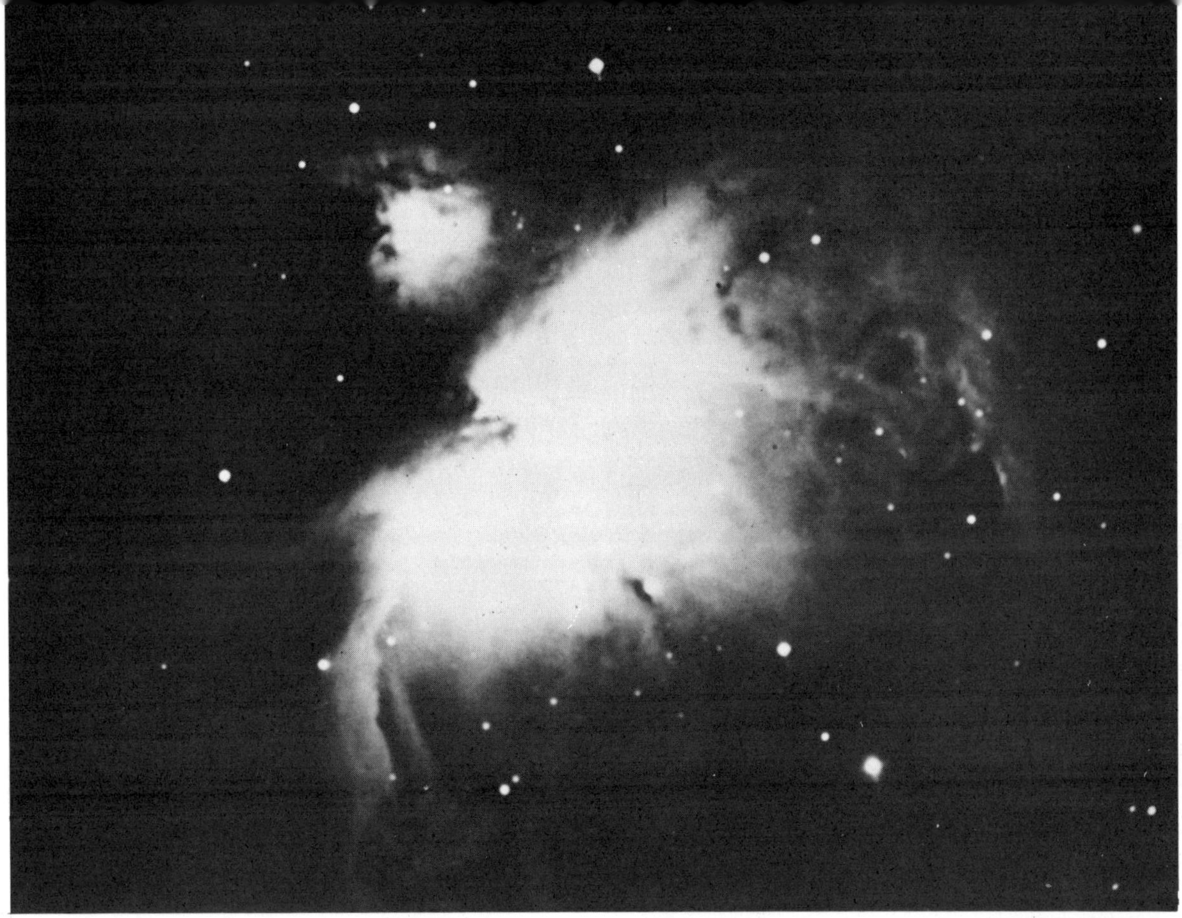

Plate 10. The Great Nebula in Orion, NGC 1976. Visible to the unaided eye as a faint patch in Orion's Sword, it is about 1300 light-years from Earth. The nebula is only about one-millionth as dense as the best vacuum scientists can produce in laboratories. *Carter Observatory*

THE GREAT NEBULA IN ORION

The fuzzy star in the middle of the handle of The Pot is not really a star at all. As revealed in telescopes, it is a vast area of gas and dust illuminated by bright stars in its interior. See Plate 10.

Analysis of radio emissions from this nebula prove that its material is in a violent state of turmoil. Certain dark patches in it could possibly be areas where stars are being born. Perhaps some of those stars will have their own planetary systems. Life might be beginning there and might make its presence known to other civilisations in space in the course of the next few thousands of millions of years.

The Orion Nebula is about 20 light-years across and it is 1300 light-years from us.

RIGEL

This star, whose official name is Beta Orionis, is bluish-white and extremely hot. It is 40,000 times more luminous than the sun and that makes it one of the great searchlights of the heavens. Only its remoteness, 850 light-years away from us, reduces its apparent brightness to seventh place in the list of the brightest stars. Its position in the constellation of Orion is shown in Fig. 17.

BETELGEUSE

Betelgeuse, Alpha Orionis, is classified as a red supergiant star and it looks orange-red in the sky. Its colour tells us that its surface is relatively cool, actually about 3000°C.

Its huge size is such that, if placed in the position of the sun, it would swallow the planets Mercury, Venus, Earth and Mars. Refer Fig. 27 page 43.

Betelgeuse varies irregularly in brightness, an effect probably caused in this case by pulsating movements within its vast bulk.

THE GREAT TRIANGLE

Sirius in the constellation Canis Major, Betelgeuse in the constellation Orion, and Procyon in the constellation Canis Minor form a conspicuous triangle whose almost equal sides measure about $1\frac{1}{4}$ handspans. The Triangle stands out boldly against a dark background devoid of other bright stars.

SIRIUS, THE BRIGHTEST OF THEM ALL

Well known to the ancients who in some countries worshipped it as a god, Sirius has always been one of the important navigational signposts of the sky. Being the brightest of all stars, it is the first to appear shortly after sunset during the summer months. Its other name is Alpha Canis Majoris.

One reason for the particular brilliance is its comparative closeness, only 8.7 light-years away. Its white colour is related to its high surface temperature of 11,000°C.

Sirius has a companion star known as Sirius B which has extraordinary properties. Its material is so dense that a man on its surface would weigh about 2000 tonnes.

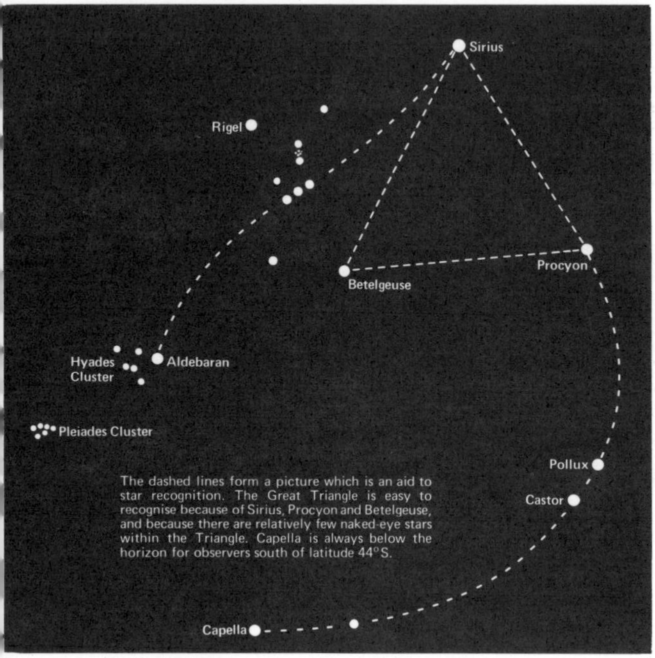

Fig. 17. Orion and its neighbours.

PROCYON

Procyon, Alpha Canis Minoris, is a brilliant yellowish star. Like Sirius, it has a small companion star which is difficult to identify even in a powerful telescope because of the glare from neighbouring Procyon.

THE TWINS

Castor and Pollux, otherwise known as Alpha Geminorum and Beta Geminorum, constitute the Twins. They both belong to the constellation Gemini and we might be excused for assuming, from their names, that Alpha Geminorum is the brighter of the two. Alpha might have been brighter in the distant past but at present it is the fainter.

A story from mythology helps the newcomer to remember these two stars. Castor and Pollux were twin sons of the god Jupiter. Pollux had the immortality of a god but Castor was mortal. When Castor was killed in battle Pollux pleaded with his father to save his dead brother from oblivion. Jupiter acted promptly, placing both of them in the sky.

CAPELLA

For southern hemisphere viewers at about Latitude 35° this yellowish star appears rather too close to the northern horizon for satisfactory observation. It never becomes visible from the southernmost areas of New Zealand. Its other name is Alpha Aurigae.

TWO FAMOUS STAR CLUSTERS

A slightly curving broken line in Fig. 17 guides the eye to a cluster of stars looking like the letter V lying on its side. This is the **Hyades** cluster whose stars, gravitationally linked, are moving as a group in space. The principal star in the cluster is Aldebaran, or Alpha Tauri, noticeably an orange-red star.

The **Pleiades** cluster is a delicately beautiful group of stars, six or seven of them detectable by people with normal eyesight. The cluster is younger than the Hyades and will preserve its identity for a considerable time, astronomically speaking.

Mythology relates that Orion, hunting one day in a forest, detected movement among the trees. He gave chase. To his amazement he came upon not animals but the seven daughters of the giant Atlas. The girls, greatly alarmed, cried out for help. Orion tried to calm them but Jupiter, the god, snatched them from the scene and installed them in the safety of the heavens, not far from Orion. To this day a popular name for the Pleiades is the Seven Sisters.

Chapter 5.

AUTUMN, WINTER AND SPRING STARS

THE SICKLE IN LEO THE LION

Think for a few minutes about the man in Fig. 11 page 8. In Chapter 2 he was looking upwards at a slanting angle towards the South Celestial Pole round which the Southern Cross and its neighbours were circling in a clockwise direction. Then, in Chapter 4, he turned round and found that Orion and its neighbours were moving in an anticlockwise direction.

To see Orion which straddles the Celestial Equator the man looked upwards, well above his horizon.

On each successive night at the same time by the clock he found that Orion had moved 1° further to the west. By March those degrees had added up to 30° and Orion was noticeably moving off stage to the west. In its place a new constellation had nearly reached mid-stage and was demanding attention.

The man is now studying Leo the Lion and he immediately notices that it is below the Celestial Equator and therefore nearer the horizon. But let us now take over from that man and look at Leo ourselves. Its appearance is indicated in Fig. 18.

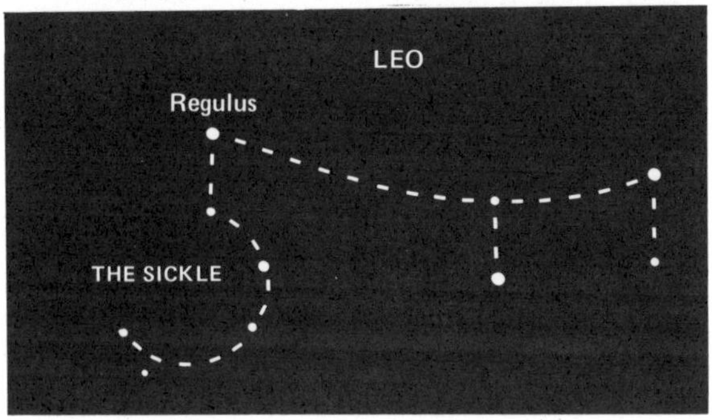

Fig. 18. The Sickle, with straight handle and curved blade, is an easily identifiable group in the constellation Leo, the Lion. (The four stars to the right of the Sickle are added as a further aid to recognition)

Six of the stars on the left of the diagram form the shape of a sickle whose handle is at the top and whose curved blade is attached below. The four stars to the right of the sickle have been included in the diagram to aid recognition. Regulus, or Alpha Leonis, is a white star at the top of the handle.

The stars of Leo do not look like a lion to us. But, add a few stars to it as shown in Fig. 19, then turn the figure upside down for a northern hemisphere view. Now the constellation is quite lion-like.

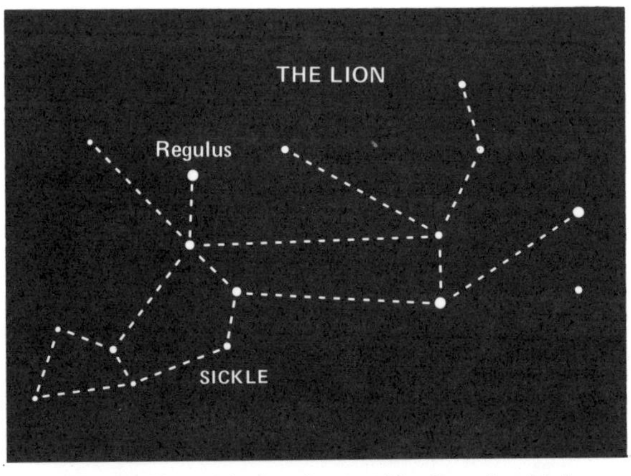

Fig. 19. Turn this drawing upside down to obtain a northern hemisphere impression of the Lion.

SPICA AND ARCTURUS

By the month of May the constellation Leo has moved over to the west and given way to Spica and Arcturus. The man in Fig. 11 discovers that Spica is just above the Celestial Equator and Arcturus is well below it.

Bright Spica lies in the constellation of Virgo the Virgin and is also called Alpha Virginis. Fig. 20 shows that above and to the left of Spica is a group of not very bright stars popularly called Spica's Spanker. Those stars are contained in the constellation Corvus, the Crow. One dictionary states that a spanker is a fore-and-aft sail set on the after side of a vessel's mizzenmast.

The yellow-orange star Arcturus, or Alpha Boötes, is classified as a red giant. It has always been a very famous star and, from its mention in the Bible, is occasionally called Job's Star. As a spectacular publicity event, light from Arcturus was used to trigger off the Chicago World's Fair in 1933.

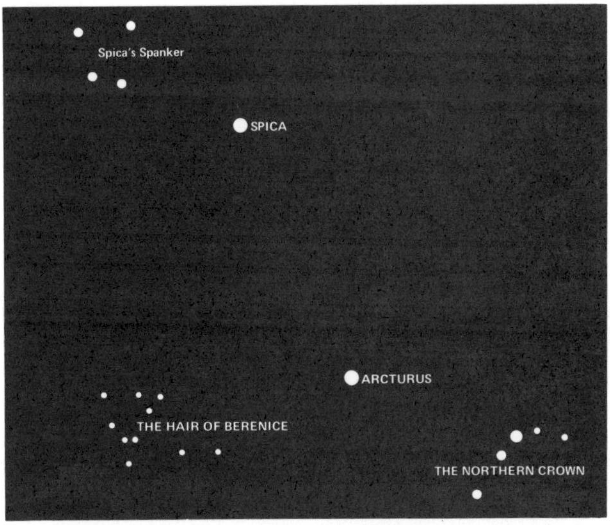

Fig. 20. Recognition of Spica and Arcturus can be confirmed by Spica's Spanker, a quadrilateral formed by stars in the constellation Corvus. Spica is in the constellation Virgo, and Arcturus is in Boötes. On clear dark nights the fainter groups of stars in Coma Berenices and Corona Borealis can be seen fairly low in the sky.

HAIR OF BERENICE

Below and to the left of Arcturus is a faint but beautiful group of stars in the constellation Coma Berenices. It has been described as resembling "gossamer spangled with dewdrops". According to legend, when Berenice's husband went off to participate in a war against the Assyrians, she vowed that if he returned safely from war she would, in gratitude, sacrifice her hair to the Goddess of Beauty. Her husband did return home and, true to her promise, Berenice cut off her hair and placed it in a temple. Later the hair was stolen. Jupiter had taken it and placed it in the sky.

THE NORTHERN CROWN

Below and to the right of Arcturus is a curved line of rather faint stars in the constellation Corona Borealis, or Northern Crown. Legend records that Bacchus presented this crown to Ariadne, daughter of the King of Crete, to comfort her after being deserted by Theseus.

SCORPIUS, THE SCORPION

While Spica and Arcturus have been occupying central areas of the northern sky, another group of stars has been rising south of east and moving into the north-east. The group bears a striking resemblance to a scorpion from which it gets its name Scorpius. By June, Scorpius is overhead and is in an upside-down attitude with its tail curving downwards instead of upwards. See Fig. 21.

By September and October, Scorpius is sinking towards the south-west horizon.

At the centre of the Scorpion's body is the prominent orange-red star Antares, flanked on each side by a faint white star. The name Antares means rival of Mars, a reference to an imagined rivalry with the red planet. Antares, or Alpha Scorpii, is a red supergiant with a huge diameter of 700 million kilometres. That makes it much larger than Betelgeuse which was mentioned in Chapter 4.

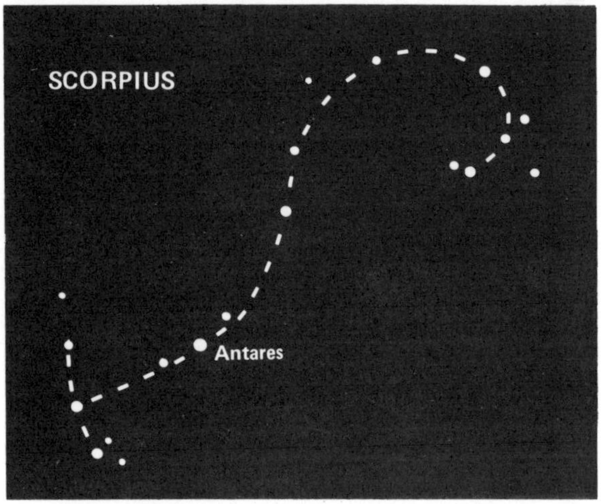

Fig. 21. Scorpius. The scorpion-like outline of the principal stars of Scorpius. In our line of sight the bright orange-red star Antares has a fainter star on each side of it.

FOMALHAUT

Shining in lonely splendour – there are few other stars nearby – Fomalhaut follows 6½ hours behind Scorpius on approximately the same course. It can be observed between August and November.

One convenient method of identifying Fomalhaut is to look for the three prominent stars which lie at about equal distances apart in an almost straight line: Canopus, Achernar, Fomalhaut. Another method, of course, is to consult an appropriate star chart such as the chart for October in the next chapter. Fomalhaut's other name is Alpha Piscis Austrini.

ALTAIR, VEGA AND DENEB

Altair, or Alpha Aquilae, comes into prominence in the northern sky after Spica and Arcturus have moved out of the way to the west. The man in Fig. 11 would find Altair below the Celestial Equator and, still further below the Equator (that is, nearer to his northern horizon), he would recognise the other two members of the Altair-Vega-Deneb triangle depicted in Fig. 22.

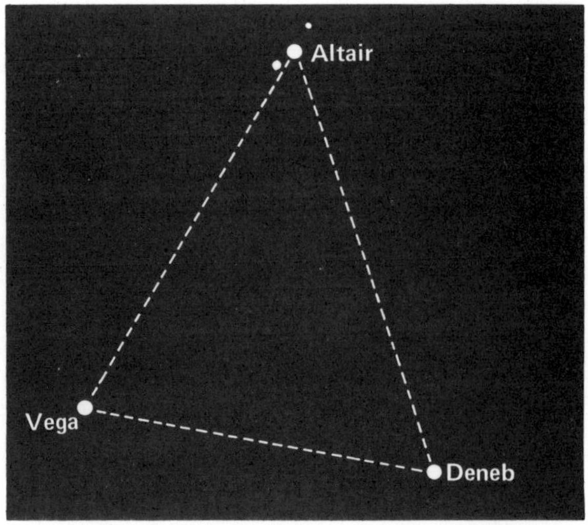

Fig. 22. Altair, Vega and Deneb. Altair, in Aquila, has a fainter star on each side of it but, unlike orange-red Antares, which also has two companion stars, Altair is white. In this bold triangle of stars the constellation Lyra is represented by Vega. Deneb belongs to the constellation Cygnus.

Aids to the recognition of Altair are its two companion stars, one on each side. In this respect it resembles Antares. But whereas Antares is orange-red, Altair is bluish-white. The companion stars of Antares and of Altair are not near those bright stars in space. In our line of sight they merely appear to be near.

The name Altair means flying vulture or eagle, which is also the interpretation of the title of the constellation to which it belongs, Aquila.

The three stars of the Altair group point downwards nearly two handspans to Vega, or Alpha Lyrae. Seen from the northern hemisphere, Vega is a very brilliant bluish star but, for southern hemisphere viewers, it is rather close to the horizon and looks less brilliant.

Our solar system is speeding through space at 19 kilometres per second in the direction of Vega. There will, of course, not be a collision because Vega is racing along in a different direction and will not be in its present spot when we arrive there.

To the right (east) of Vega, and nearer the horizon, is Deneb, which in its own right is one of the most luminous stars known. Only its great distance away, 1500 light-years, keeps its apparent brightness at the comfortable level we see.

Observers in the far south of New Zealand do not see Deneb at all. If the man in Fig. 11 wants to obtain the best views of Vega and Deneb he should travel into the northern hemisphere (downwards in the diagram).

THE SQUARE OF PEGASUS

By October, the Altair-Vega-Deneb triangle has given place to the Square of Pegasus which is shown in Fig. 23. The "square" is distinctive but the stars are comparatively faint.

The star Alpheratz is in the constellation Andromeda. The other three stars in the "square" belong to the constellation Pegasus.

In mythology, Pegasus was a wonderful winged horse which possessed immortality. With its magic powers it was able to elude any man trying to catch it – that is, until one man, Perseus, not only caught the horse but also managed to ride it on many journeys of adventure. One day Perseus on his horse discovered the maiden Andromeda chained to a rock and unable to escape from a monster threatening to devour her. With great gallantry Perseus rescued the maiden and later married her. Perseus, Andromeda and Pegasus lived happy lives and were eventually placed in the sky.

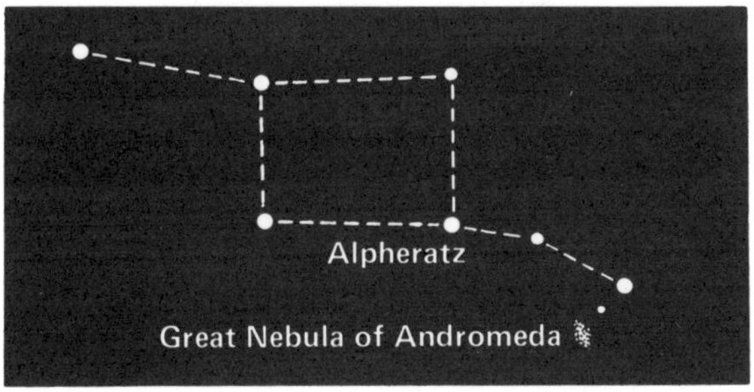

Fig. 23. The Square of Pegasus. Three stars of the square lie in the constellation Pegasus but Alpheratz is in the constellation Andromeda.

Fig. 24. Pegasus, the winged horse of mythology.

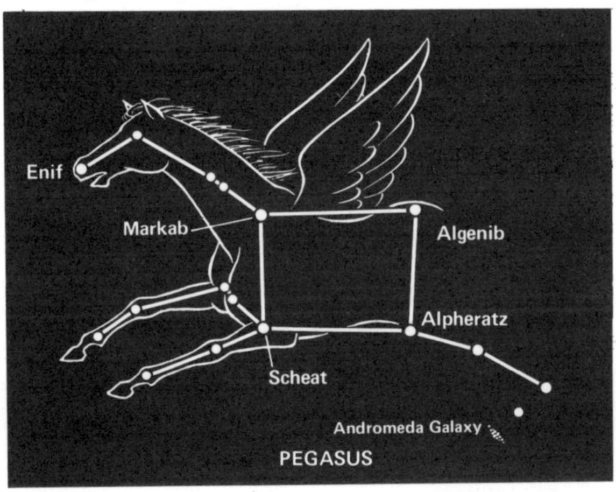

THE ANDROMEDA GALAXY

This grand object below the Square of Pegasus is too close to the northern horizon for favourable viewing from northern parts of New Zealand. From the far south of New Zealand it cannot be seen at all, being below the horizon altogether. However, the galaxy is well placed for observation from most parts of Australia.

The Andromeda galaxy (Plate 35 page 58) is one of the greatest natural objects studied by man. Visible to the naked eye as a dim patch of light, it blossoms into a huge collection of stars and luminous areas when viewed through a powerful telescope or when a long-exposure photograph of it is examined. It is 2.2 million light-years from us and its diameter is about 200,000 light-years. In the picture can be seen Andromeda's two satellite galaxies named M32 and NGC205.

PLANISPHERE FOR THE NORTHERN HEMISPHERE

The planisphere pictured in Plate 8, Chapter 3, was designed for observers located at or near latitude 35° South and it is useful within the range 30°S to 40°S.

If, however, we wish to become knowledgeable about the sky as seen from the northern hemisphere we can purchase a Philips' Planisphere designed for latitude 42° North, with a useful range of 37°N to 47°N.

Plate 11. A planisphere for identifying stars as seen from the northern hemisphere.

Plate 12. Members of a pupil-and-parent class studying a star chart at Auckland Observatory.

A. L. Rounthwaite

V. THE PRINCIPAL STARS AT A GLANCE

Chapter 6.

MONTHLY STAR CHARTS

The following 12 charts show instantly which stars are visible during the evenings of each month of the year.

HOW TO USE THE CHARTS

Select the chart for the required month and find from it which stars should be seen towards the south and towards the north. Then, using an electric torch if desired, step out of doors and compare the chart with the real sky. If necessary, rotate the chart clockwise or anticlockwise until the positions of the stars in the chart match those of the actual stars.

The reason why the stars change their positions during any one night, and from night to night, have been explained in Chapter 3.

TIMES STATED ON THE CHARTS

The charts show the positions of stars at certain standard (or clock) times. However, New Zealand viewers should add 30 minutes to those times. This is necessary because New Zealand Standard Time has, incorporated in it, 30 minutes which was added in 1928 as a Summer Time measure. The Standard Time Act of 1945 provided that the time corresponding to Longitude 180°, which is 12 hours in advance of Greenwich Mean Time, be adopted as New Zealand Standard Time.

DAYLIGHT TIME

Many countries operate Daylight Time over a period of each year. New Zealand Daylight Time is 13 hours ahead of Greenwich Mean Time (Universal Time) and it is in use from 2 am New Zealand Standard Time on the last Sunday in October until 2 am New Zealand Standard Time on the first Sunday in March.

Thus, during that period, people in New Zealand should add one hour to New Zealand Standard Time i.e. a total of 1½ hours to the standard times on the charts.

JANUARY

Viewing time (standard): mid-month at 10.30 pm

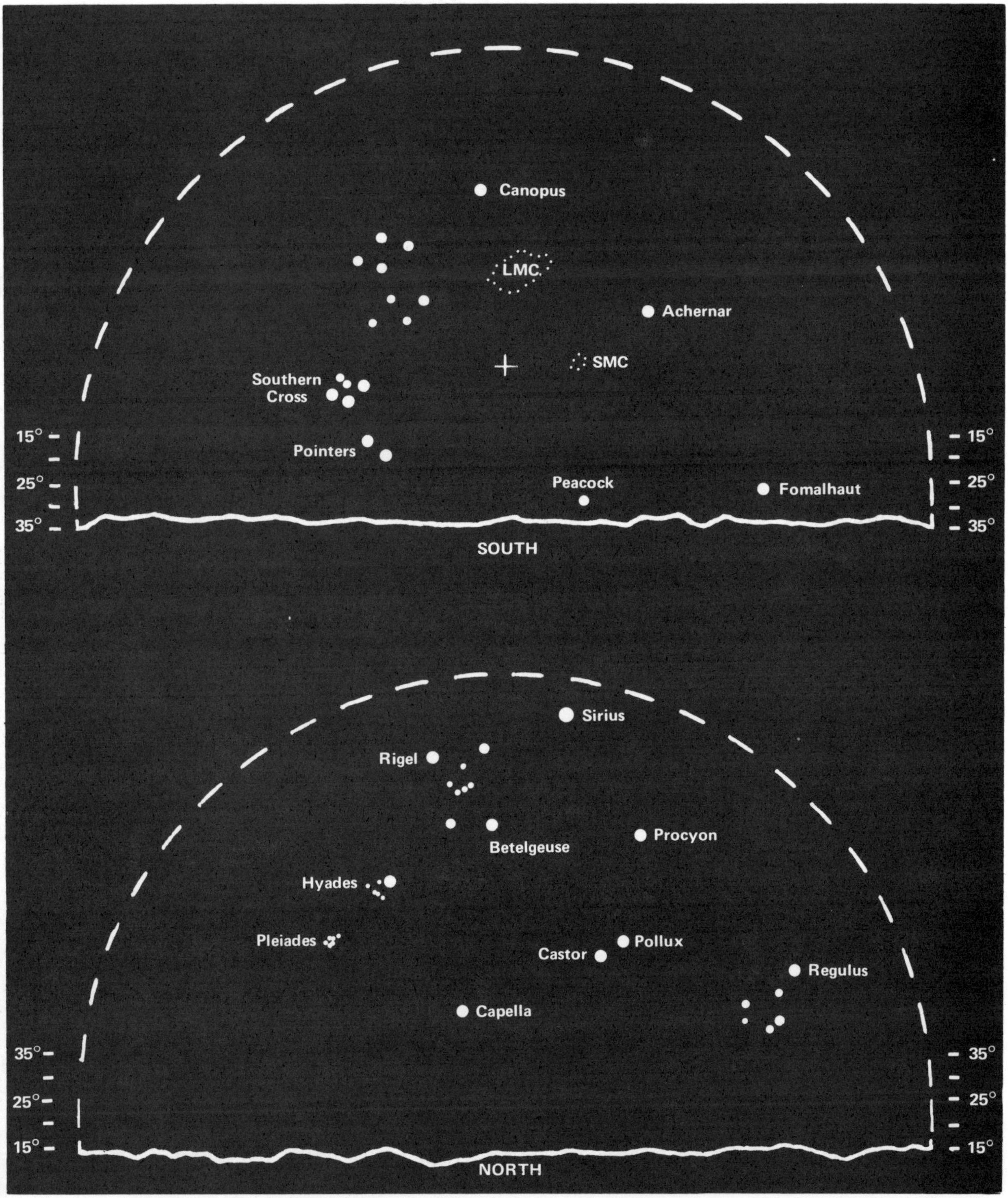

Other viewing times (standard): Jan 1 at 11.30 pm, Jan 30 at 9.30 pm

For New Zealand Standard time see Chapter 6, page 18.

FEBRUARY

Viewing time (standard): mid-month at 10.30 pm

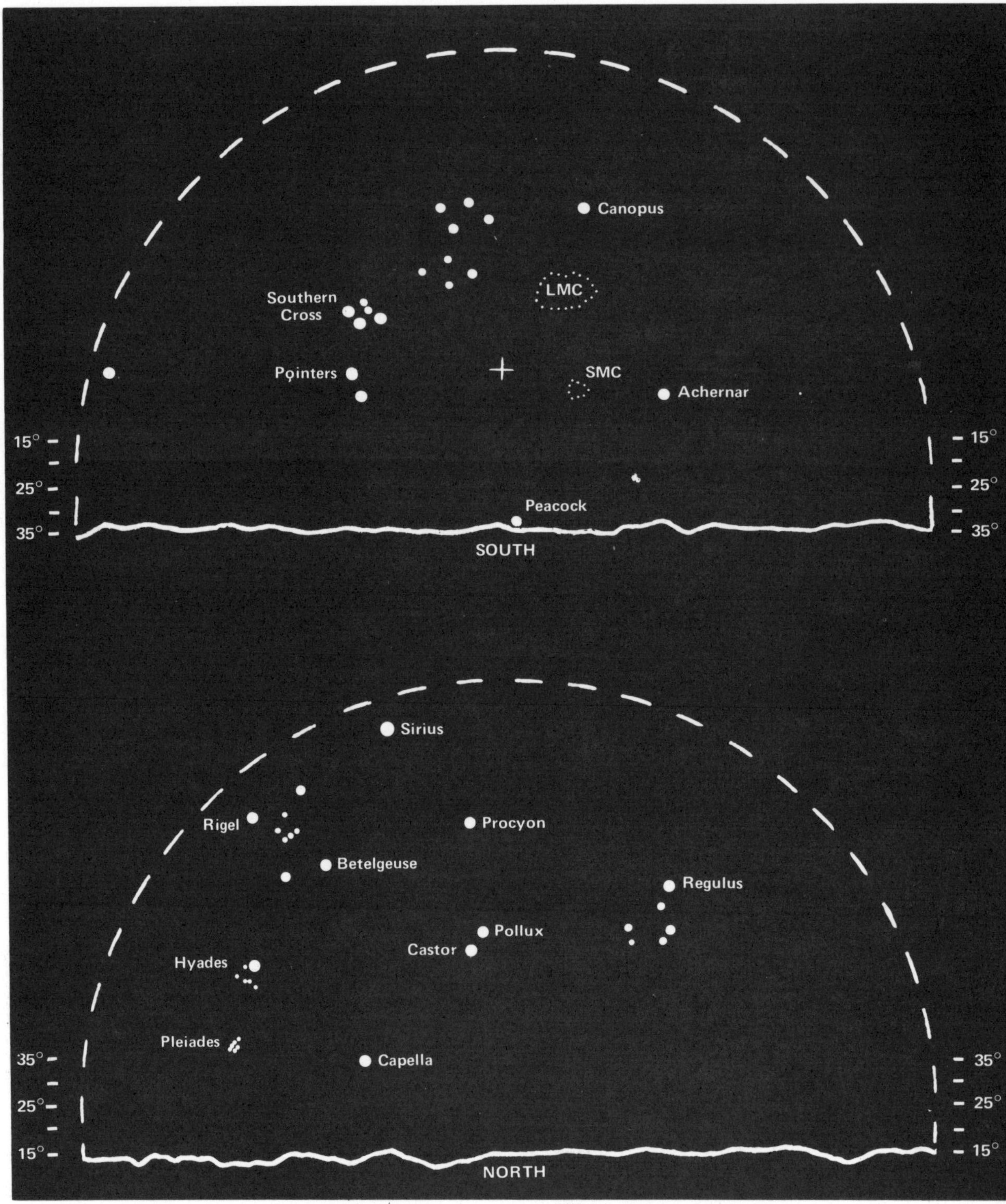

Other viewing times (standard): Feb 1 at 11.30 pm, Feb 28 at 9.30 pm

For New Zealand standard time see Chapter 6, page 18.

MARCH

Viewing time (standard): mid-month at 10.30 pm

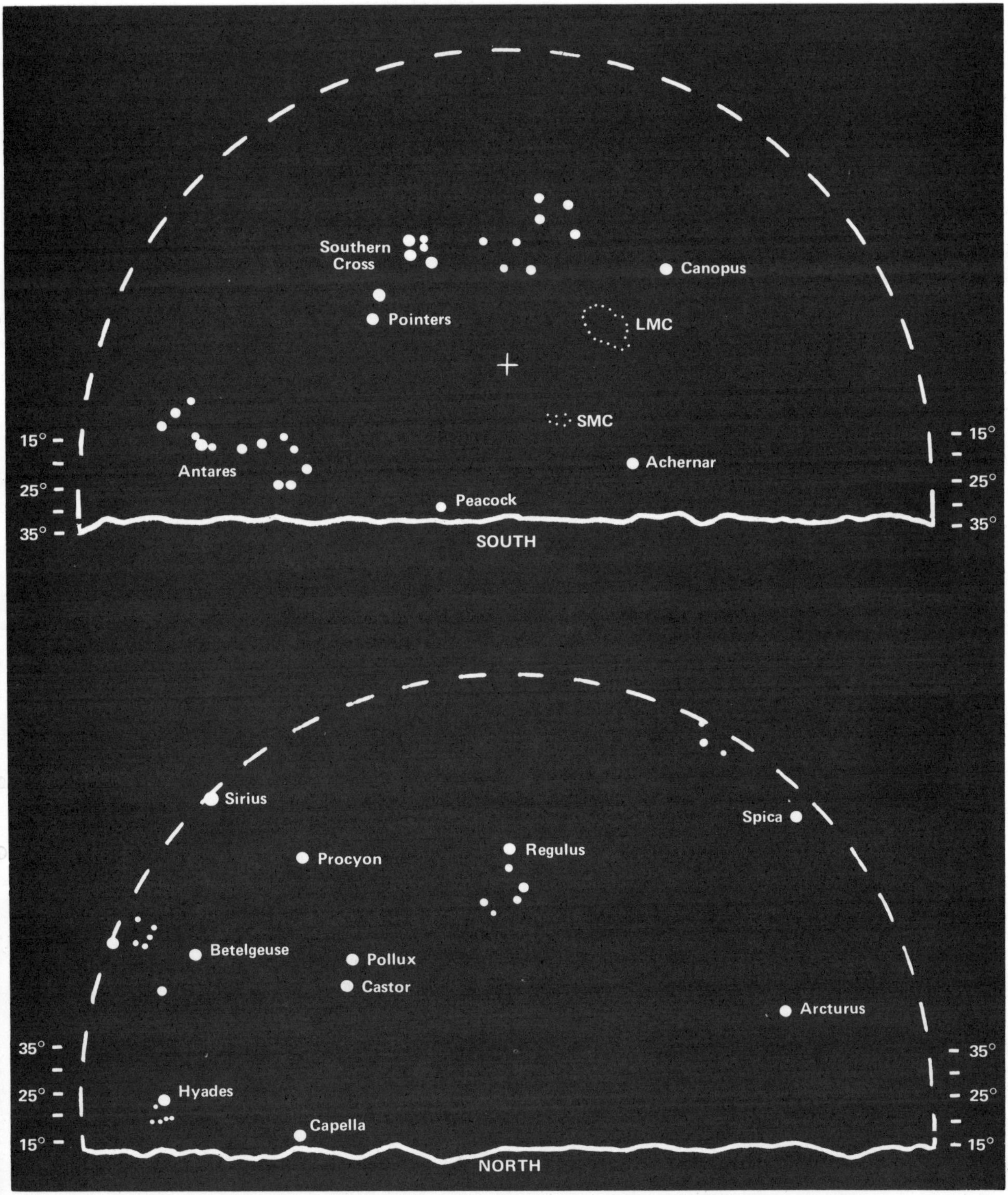

Other viewing times (standard): Mar 1 at 11.30 pm, Mar 30 at 9.30 pm
For New Zealand standard time see Chapter 6, page 18.

APRIL

Viewing time (standard): mid-month at 10.30 pm

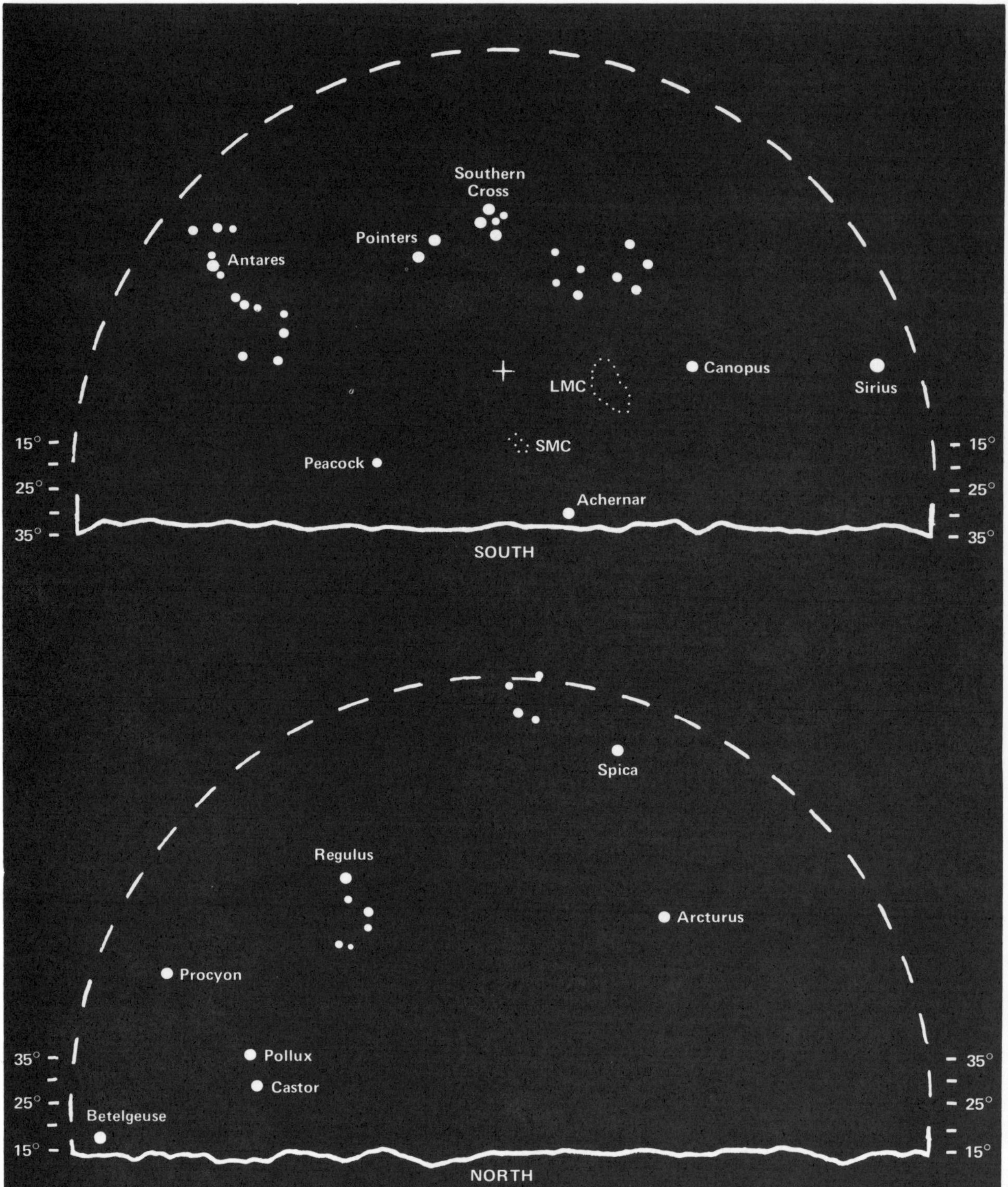

Other viewing times (standard): Apr 1 at 11.30 pm, Apr 30 at 9.30 pm

For New Zealand standard time see Chapter 6, page 18.

MAY

Viewing time (standard): mid-month at 10.30 pm

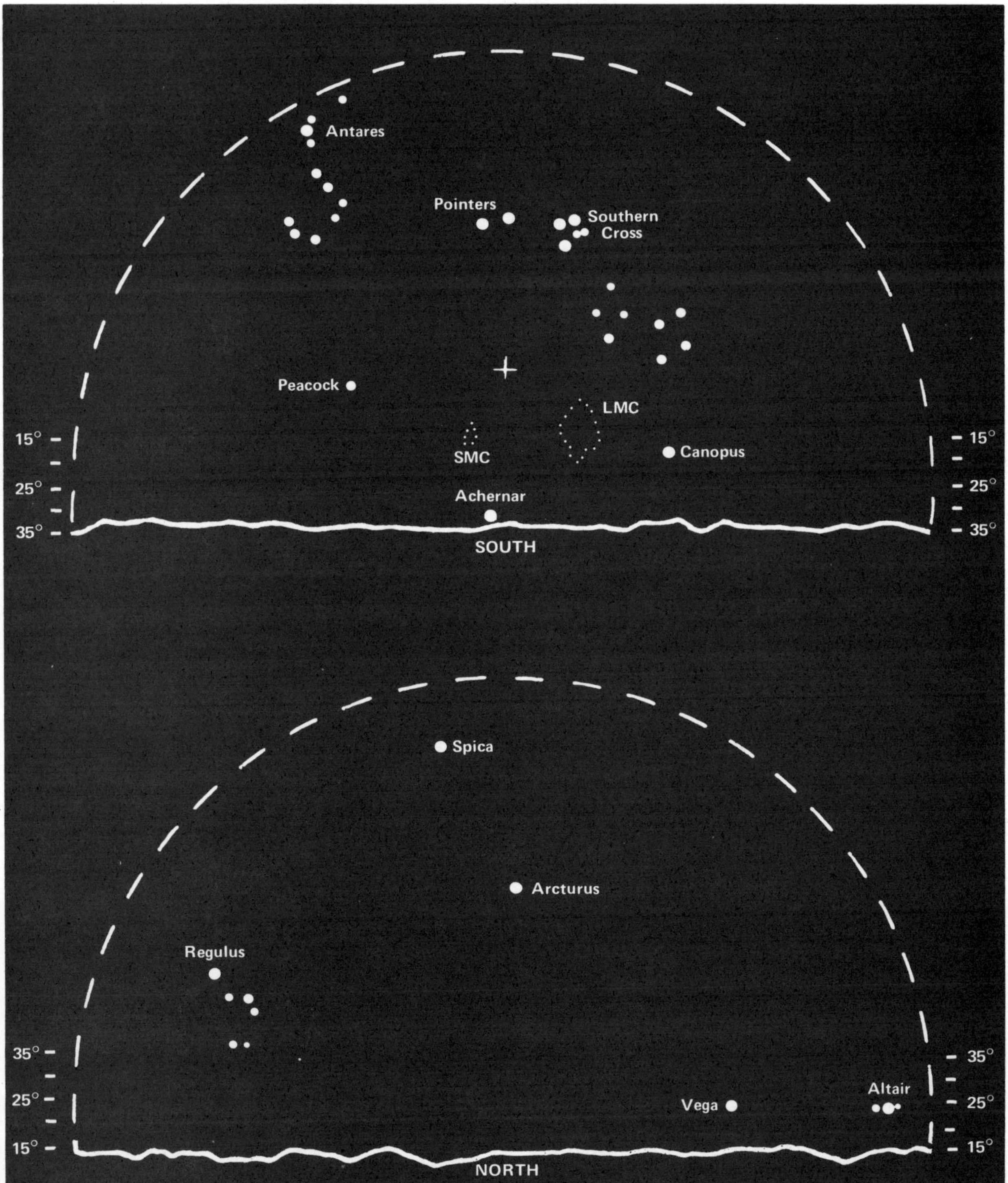

Other viewing times (standard): May 1 at 11.30 pm, May 30 at 9.30 pm
For New Zealand standard time see Chapter 6, page 18.

JUNE

Viewing time (standard): mid-month at 10.30 pm

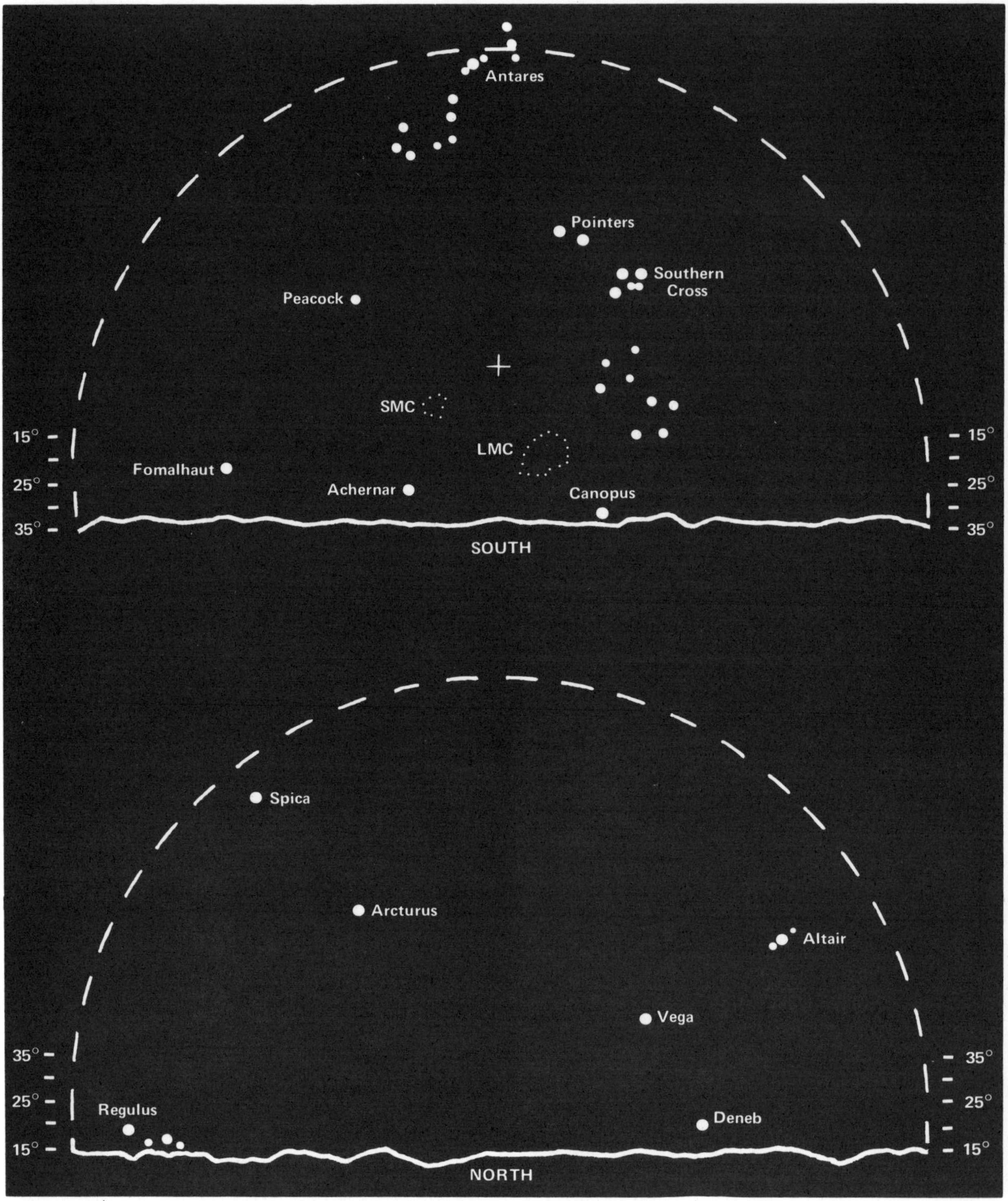

Other viewing times (standard): June 1 at 11.30 pm, June 30 at 9.30 pm
For New Zealand standard time see Chapter 6, page 18.

JULY

Viewing time (standard): mid-month at 10.30 pm

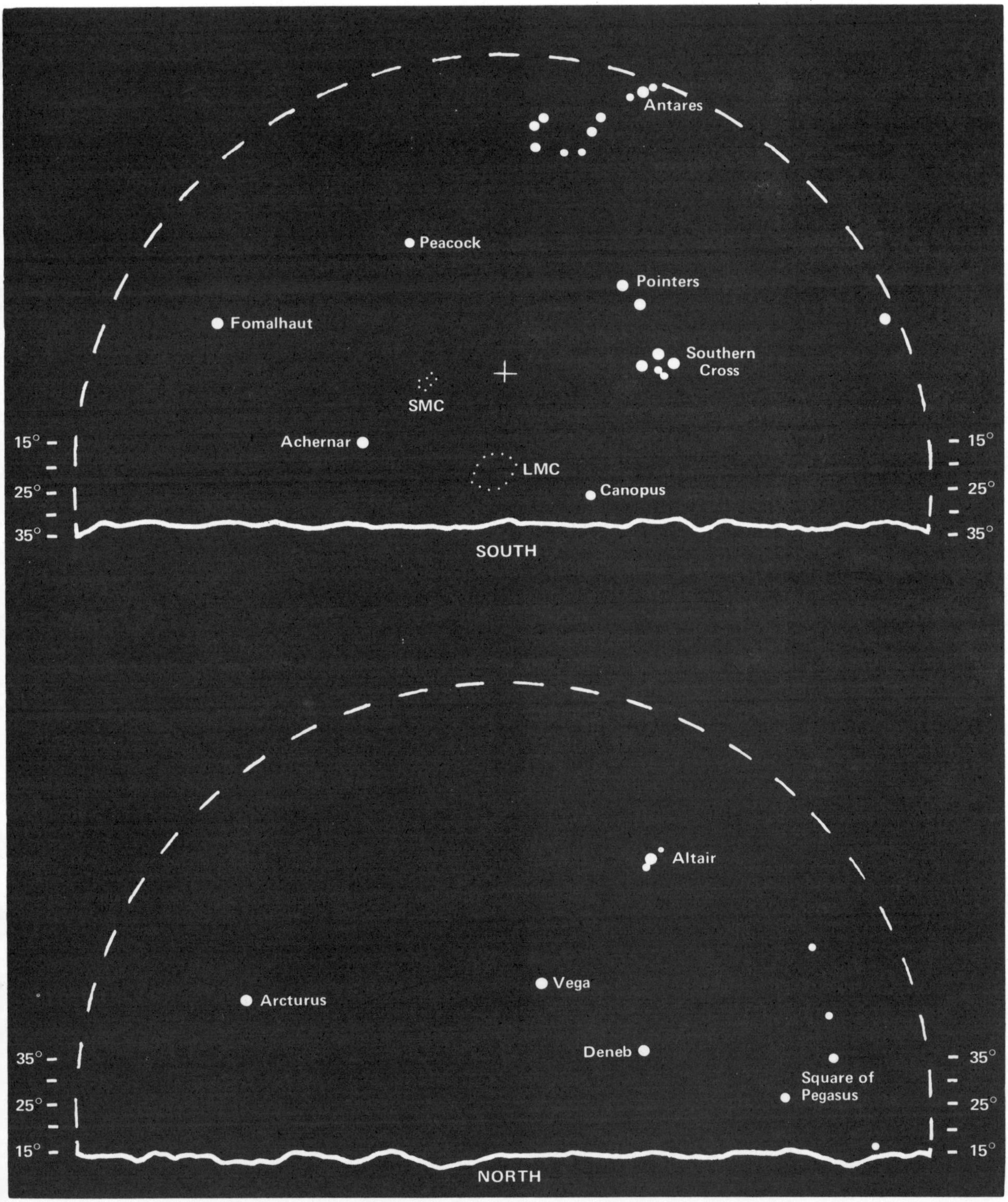

Other viewing times (standard): July 1 at 11.30 pm, July 30 at 9.30 pm

For New Zealand standard time see Chapter 6, page 18.

AUGUST

Viewing time (standard): mid-month at 10.30 pm

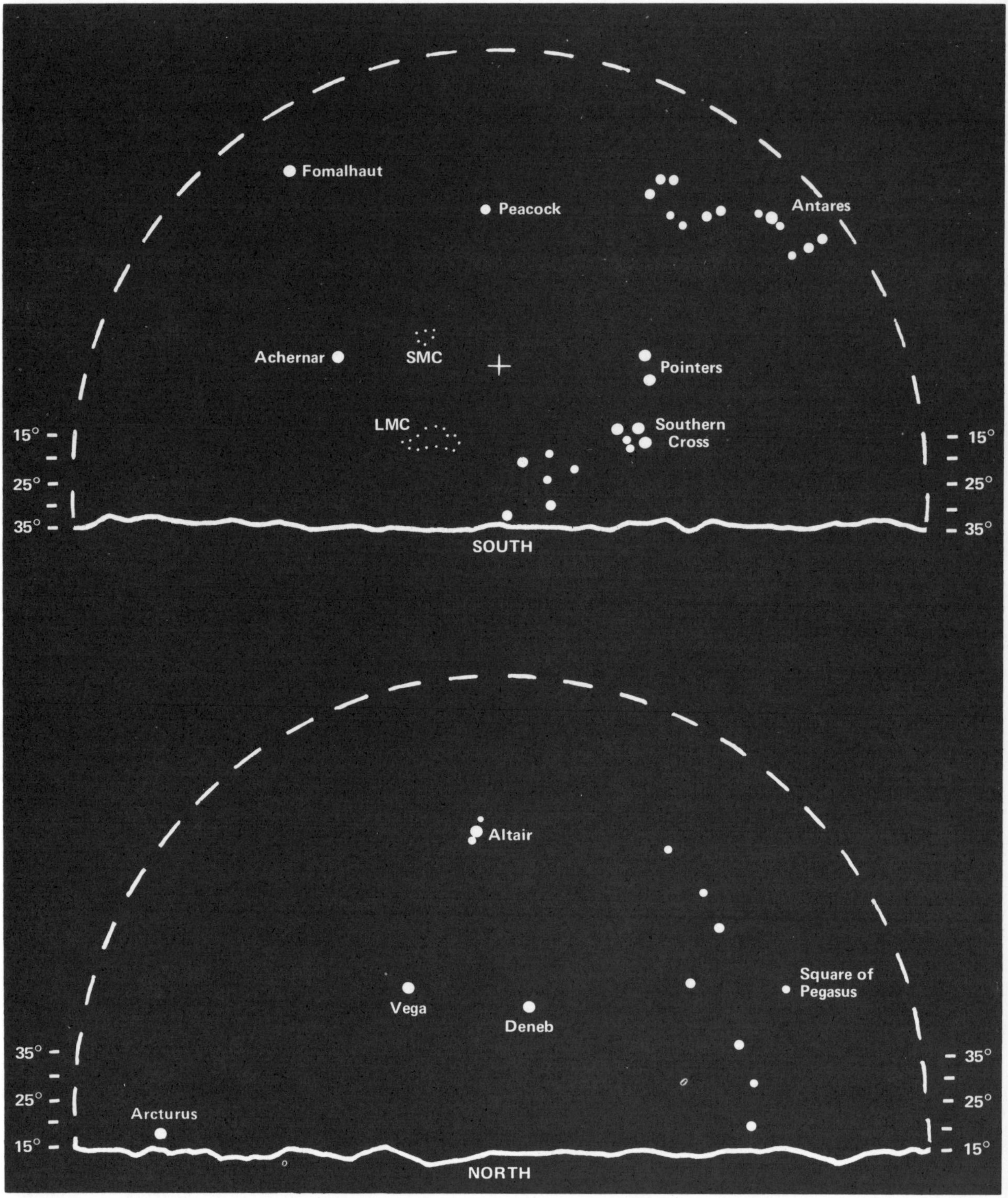

Other viewing times (standard): Aug 1 at 11.30 pm, Aug 30 at 9.30 pm

For New Zealand standard time see Chapter 6, page 18.

SEPTEMBER

Viewing time (standard): mid-month at 10.30 pm

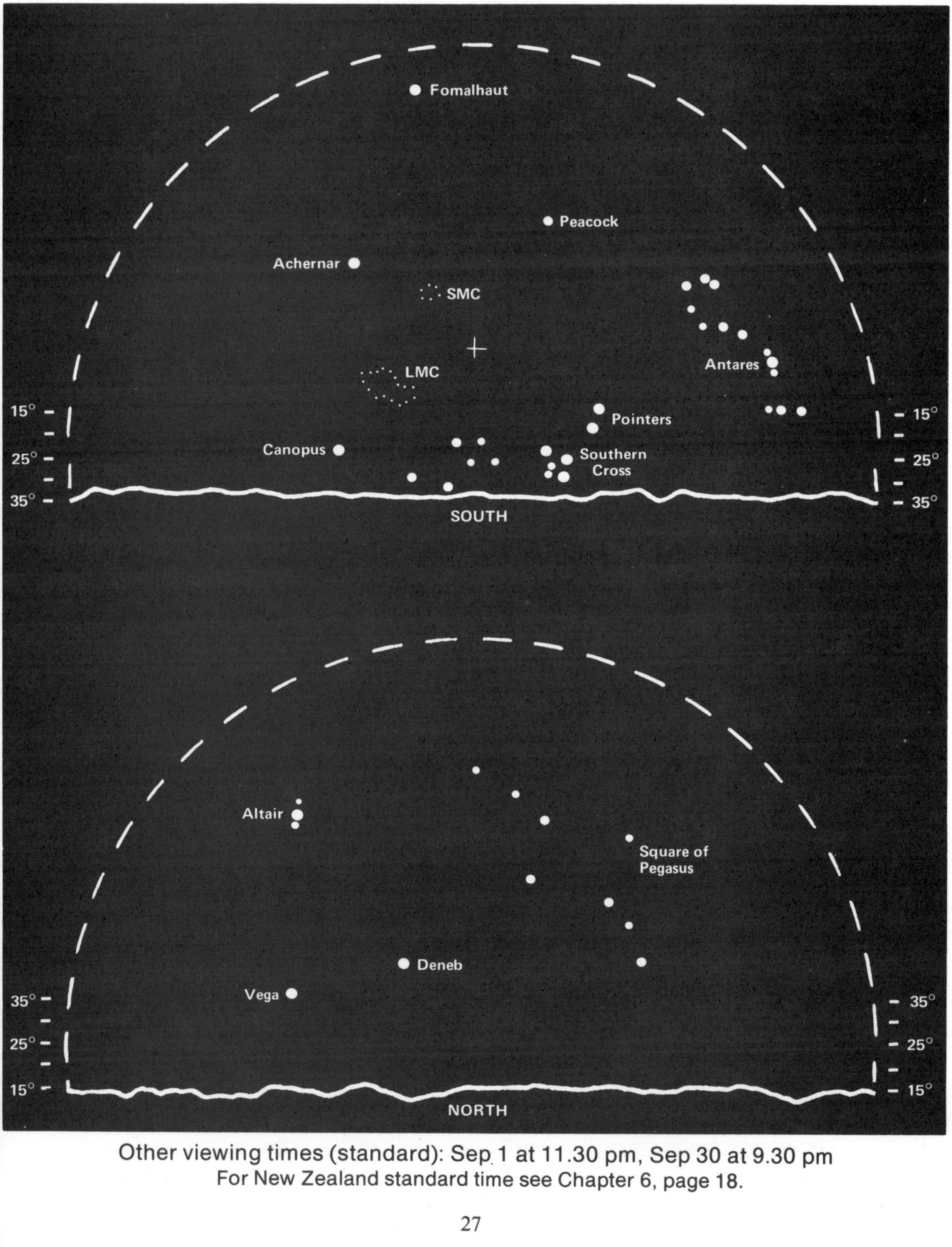

Other viewing times (standard): Sep 1 at 11.30 pm, Sep 30 at 9.30 pm
For New Zealand standard time see Chapter 6, page 18.

OCTOBER

Viewing time (standard): mid-month at 10.30 pm

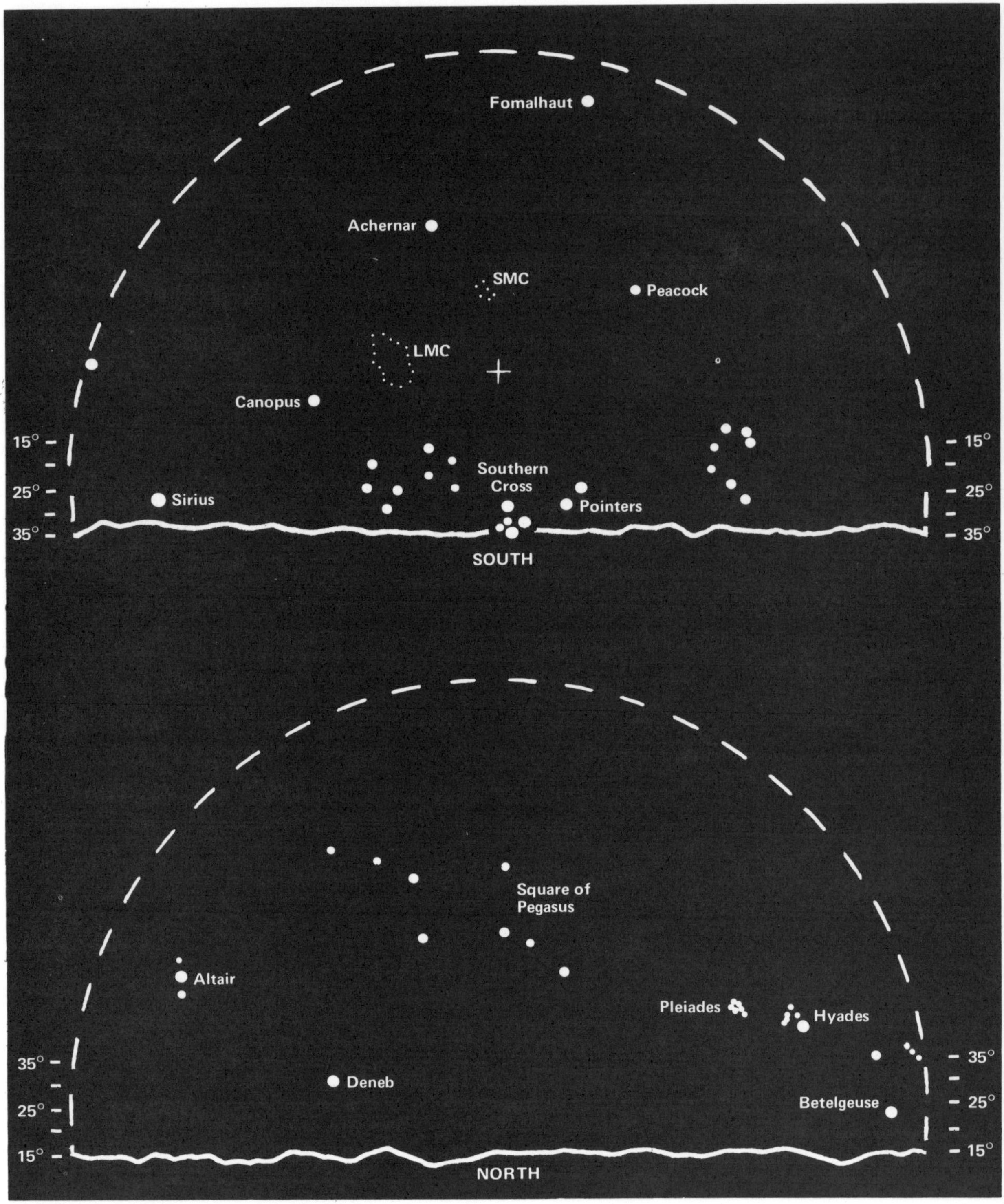

Other viewing times (standard): Oct 1 at 11.30 pm, Oct 30 at 9.30 pm

For New Zealand standard time see Chapter 6, page 18.

NOVEMBER

Viewing time (standard): mid-month at 10.30 pm

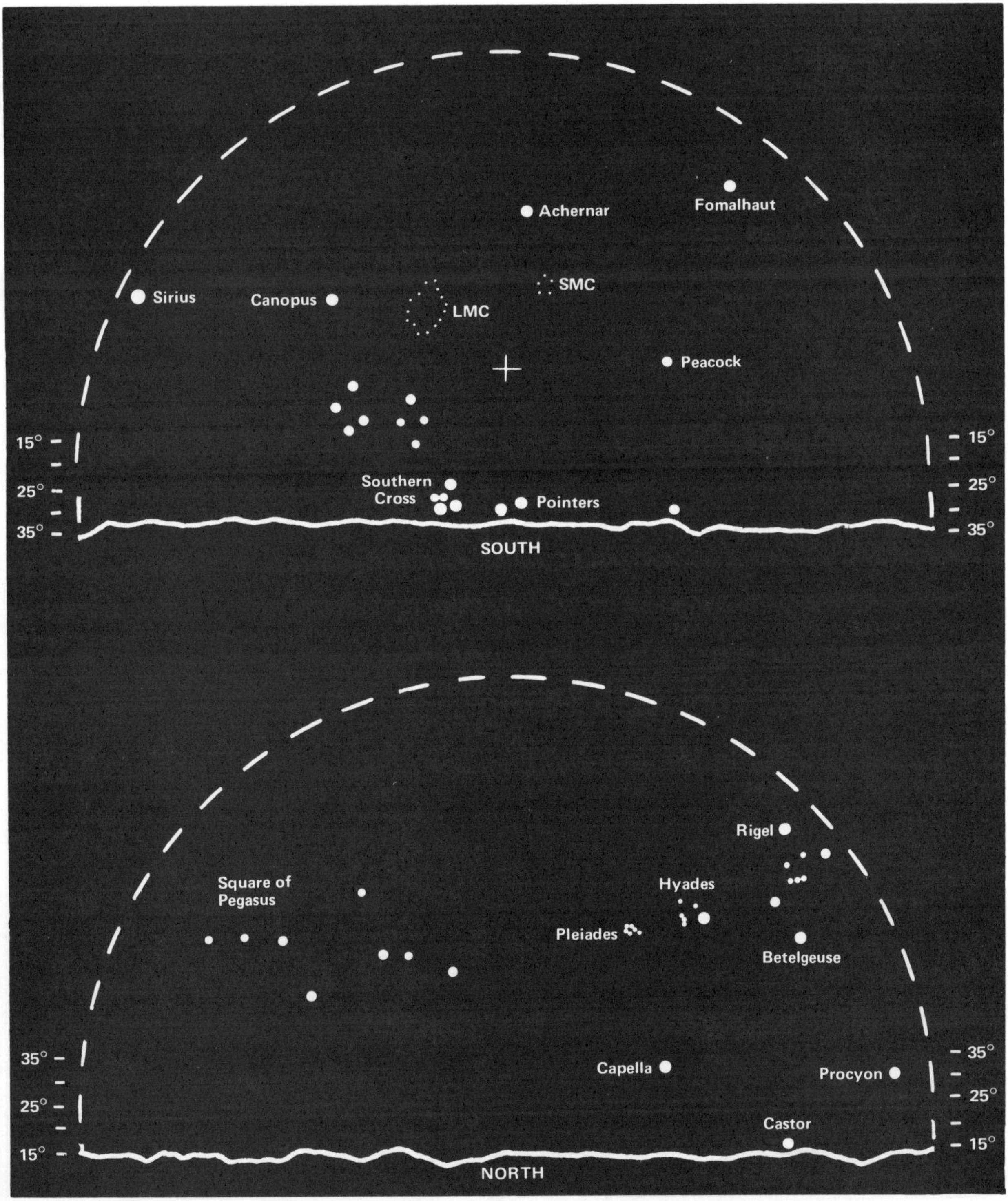

Other viewing times (standard): Nov 1 at 11.30 pm, Nov 30 at 9.30 pm

For New Zealand standard time see Chapter 6, page 18.

DECEMBER

Viewing time (standard): mid-month at 10.30 pm

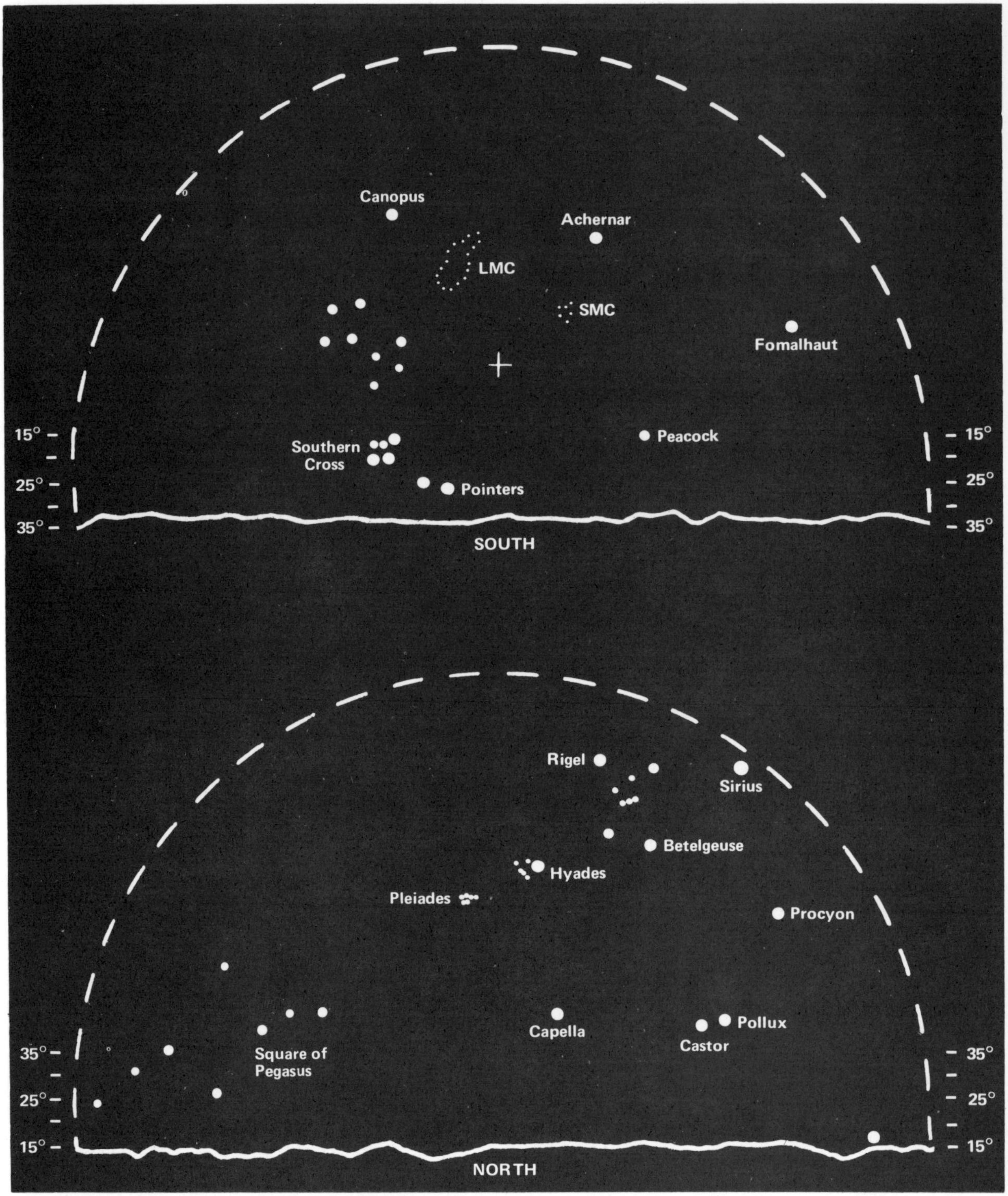

Other viewing times (standard): Dec 1 at 11.30 pm, Dec 30 at 9.30 pm

For New Zealand standard time see Chapter 6, page 18.

VI. SEARCHING THE SKY WITH BINOCULARS AND TELESCOPE

Chapter 7.

STARS, NEBULAE AND GALAXIES

Views of the night sky become fascinatingly enriched when examined through binoculars or a telescope of good quality. (Refer to Appendix A for hints on how to choose them.)

These instruments can bring into instant prominence many objects which are faint or invisible to the naked eye. Some individual stars are revealed as double, triple or multiple stars. Certain faint patches blossom out as beautiful clusters of stars, or vast clouds of dust and gas, or mighty galaxies crowded with a variety of celestial wonders. A few of the planets display surface markings, rings (in the case of Saturn) or attendant moons.

Plate 13. Astronomy students receiving instruction in aiming a 102 mm refractor telescope with equatorial mounting. To ensure eye safety during daytime the dust cap has been placed in position on the upper end of the telescope. Author far right.

A. L. Rounthwaite

HOW TO FIND THEM

Some of the most famous of such objects are described, together with simple directions for finding them in the sky, in this chapter. For identification purposes Whole Year Charts I–V are provided at the end of the chapter.

Suppose, as an example, you wish to locate the Great Looped Nebula referred to in the SUMMER section on page 32. Proceed as follows:

1. **Matching the chart with the sky:** The text states that this object is within the Large Magellanic Cloud (No. 2 in Chart I). Find this Magellanic

Cloud in the chart and in the sky, rotating the chart if necessary until it matches the sky.

2. **Naked eye:** With the naked eye, locate the Great Looped Nebula. Become familiar with the general pattern of the stars in the vicinity.

3. **Binoculars:** Examine the Looped Nebula, noting that it has become enlarged and brightened in the binoculars. Again explore the general region.

4. **Telescope:** Locate the object in the telescope's "finder" and then in the main telescope. Obtain a mental impression of the overall shape of the Looped Nebula as well as of constituent and neighbouring patches. Note the pattern of surrounding stars and check that they correspond with those on Plate 14. By this time you will be sure you have correctly identified the Looped Nebula and nearby areas of the sky.

DECLINATION AND RIGHT ASCENSION

A knowledge of these topics is not required for locating the objects covered in this chapter. However, they are explained on page 37 to provide added understanding of official methods for specifying the positions of objects on the celestial sphere.

SUMMER

December, **January,** February

THE GREAT LOOPED NEBULA

On any clear, dark night look for this object with the unaided eye, recognising it as a bright patch within the large Magellanic Cloud (No. 2 in Chart I).

Binoculars will make it more distinct and a small telescope will improve that view. In a large telescope its full glory is revealed in the form of a vast nebula surrounding the hot star 30 Doradus.

This nebula is also known as NGC 2070 which refers to its number 2070 in the **New General Catalogue of Nebulae and Clusters.** Another, more popular, name is the Tarantula Nebula because its shape resembles that of the tarantula spider.

THE LARGE MAGELLANIC CLOUD

This magnificent object is really a galaxy as explained in Chapter 2.

Binoculars are recommended for leisurely inspection of many of its individual stars, clusters, nebulae and its infinite variety of random star patterns. In Chart I it is labelled LMC.

Another name for it is Nubecula Major. Nubecula means Little Cloud and Major means Greater.

RIGEL, A HOT BLUISH-WHITE BINARY

Notice how brilliantly this star shines through even a small telescope. Rigel, mentioned on page 13, is a binary which is one type of double star (No. 4 in Chart II).

A **double star** is an object which appears to the naked eye as one star but which shows up in a telescope as two stars. The two component stars may be very distant from each other though appearing to be close in our line of sight. Alternatively two stars may be relatively close and connected by gravitational pull. In that case they are swinging round a common centre of gravity and the pair is called a **binary.**

The period of time taken by the components of a binary to swing round the centre of gravity ranges in different cases from a day or so to many centuries.

BETELGEUSE, THE RED SUPERGIANT

In a telescope Betelgeuse (see also page 13) provides a startling contrast to Rigel in colour. It varies irregularly in brightness as it pulsates in a state of imbalance between the pressures of radiation outwards and gravity inwards (No. 7 in Chart II).

SIGMA ORIONIS, THE FOUR-STAR GROUP

The number of stars detectable in this group will depend on the size and performance of your telescope. Look for a bright star with three faint companions (No. 6 in Chart II).

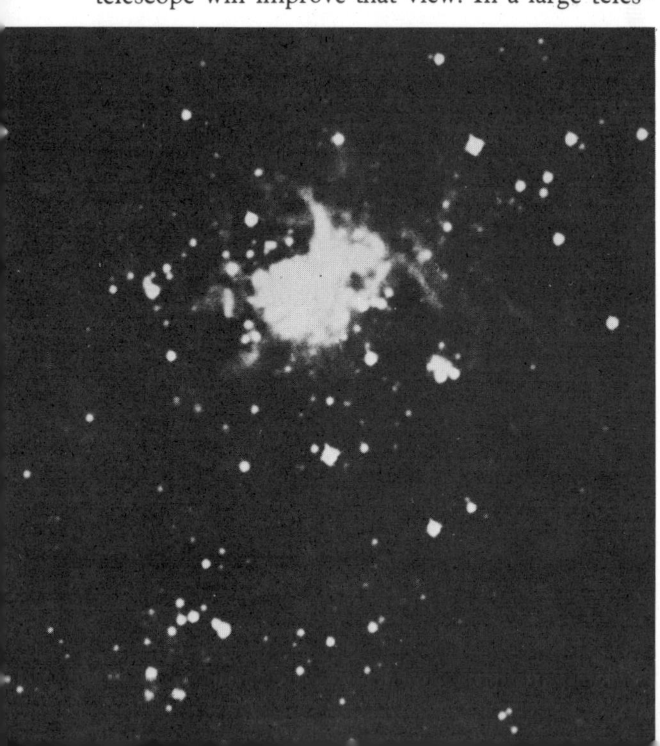

Plate 14. The Great Looped Nebula, NGC 2070, visible to the naked eye in the Large Magellanic Cloud, is sometimes called the Tarantula Nebula. It is one of the largest and brightest nebulae known.

Carter Observatory

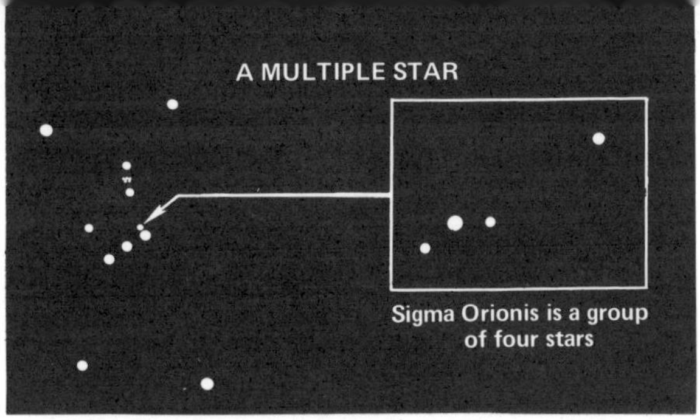

Fig. 25. The four stars of Sigma Orionis are gravitationally linked. Each star follows an orbit whose shape is constantly changing because of the differing gravitational pulls between itself and other members of the group. The group as a whole, however, remains virtually intact as it journeys through space.

CASTOR, A BINARY

Brilliant Castor is a binary, each component of which is itself a binary. The main components are now too close in our line of sight for identification in a small telescope (No. 9 in Chart II.)

PRAESEPE, THE BEEHIVE CLUSTER

About four moon-widths below and to the left (west) of Delta Cancri look for the Praesepe cluster appearing as a white patch to the naked eye but as a group of stars in binoculars. In this cluster is an attractive orange star with an orange-red companion (No. 10 in Chart III).

THE HYADES CLUSTER AND THE PLEIADES CLUSTER

These beautiful objects, introduced in Chapter 4, are well displayed in binoculars. A telescope spreads their individual stars so far apart that the unity of each cluster is lost. (Labelled objects in Chart II.)

AUTUMN

March, **April,** May

THE ETA CARINAE NEBULA, NGC 3372

This is one of the truly great sights in our southern skies. It is a vast region containing the remnants of a mighty explosion in the star Eta Carinae. That star, seen by Edmond Halley in 1677 as an object faintly visible to the naked eye, experienced many fluctuations in brightness during ensuing years. By the 1840s it was outshining all other stars except Sirius. Later it became dim and at the present time is just below naked-eye visibility.

The NGC 3372 nebula surrounding it is impressive even in small telescopes but large telescopes present it as an extraordinarily beautiful object. It is sometimes called the Keyhole Nebula in reference to a keyhole-shaped area within it (No. 5 in Chart I).

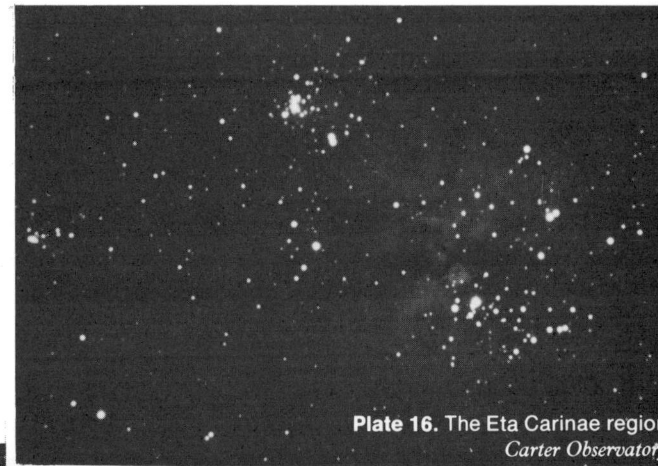

Plate 16. The Eta Carinae regio
Carter Observator

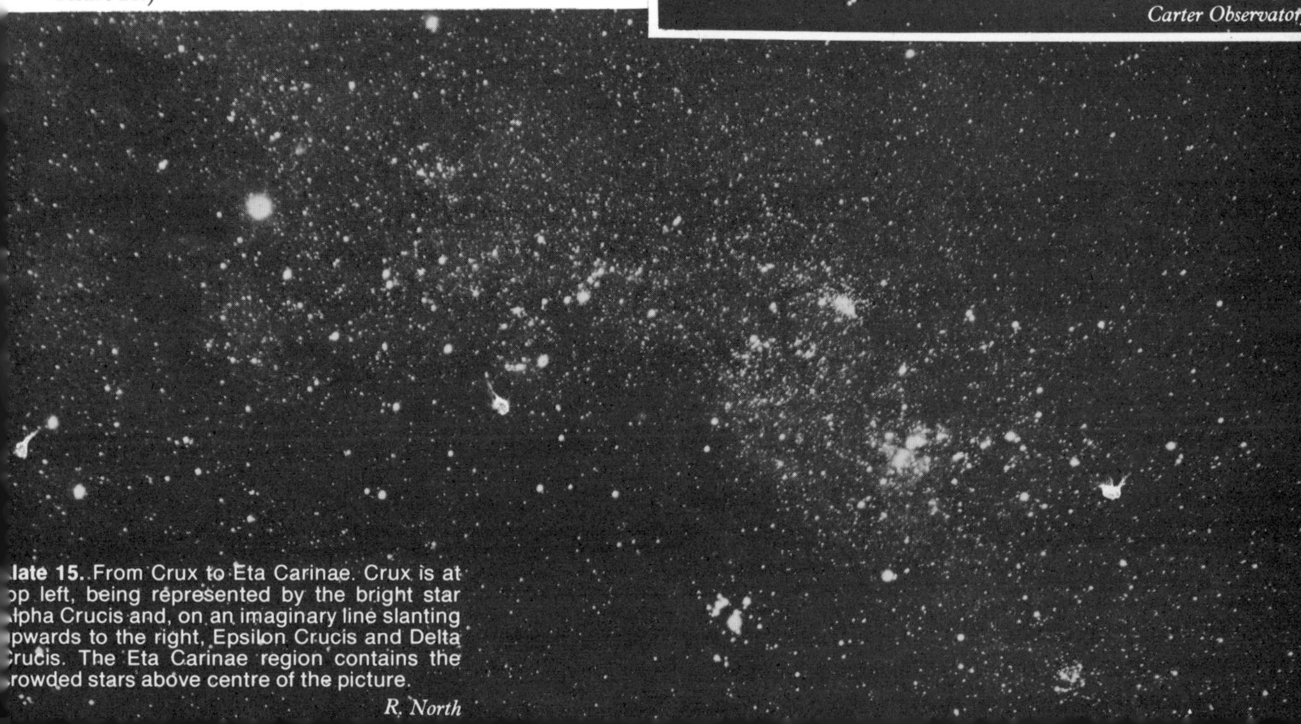

late 15. From Crux to Eta Carinae. Crux is at op left, being represented by the bright star lpha Crucis and, on an imaginary line slanting pwards to the right, Epsilon Crucis and Delta rucis. The Eta Carinae region contains the rowded stars above centre of the picture.
R. North

NGC 2516 CLUSTER NEAR THE FALSE CROSS

With the unaided eye NGC 2516 can be detected near the star Epsilon Carinae in the False Cross. Binoculars improve the view. (No. 4 in Chart I.)

ALPHA CRUCIS, A BRILLIANT DOUBLE

Both of its two bluish components are visible in a small telescope (No. 7 in Chart I).

THE JEWEL CASKET

One of the favourite telescopic views near the Southern Cross is the Jewel Casket, the cluster NGC 4755. It is a brilliant group surrounding the red star Kappa Crucis, which is close to bright Beta Crucis. It can be seen in small telescopes but becomes a glorious sight in a large telescope such as is operated at public observatories (No. 9 in Chart I.)

COMA BERENICES (HAIR OF BERENICE)

Find this attractive group of stars with the naked eye and, with a telescope, look for hazy objects which are distant galaxies. (Hair of Berenice is labelled in Chart III.)

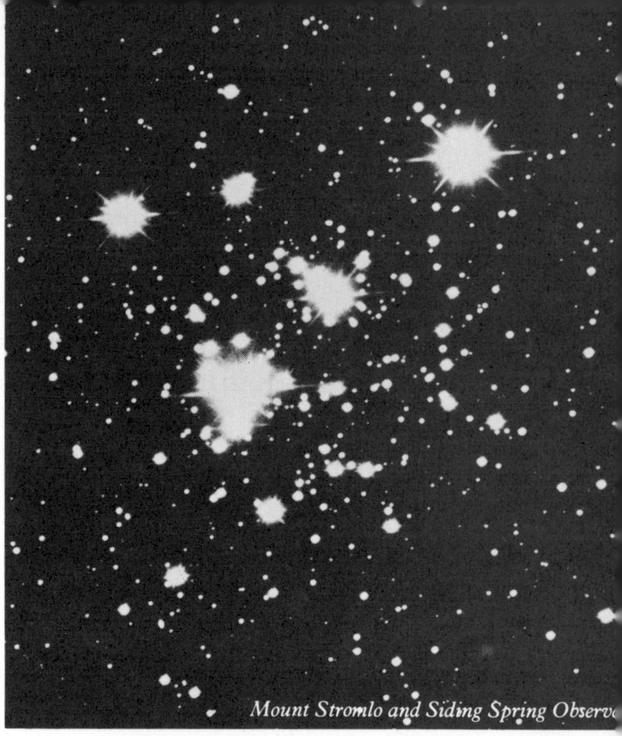

Mount Stromlo and Siding Spring Observ

Plate 17. The Jewel Casket, a beautiful cluster of more than 100 stars of many colours. At the centre is the red star Kappa Crucis.

Plate 18. The globular cluster Omega Centauri. Consisting of thousands of stars, it is one of the noblest spectacles in the southern sky.

Mount Stromlo and Siding Spring Observatory

WINTER

June, **July,** August

OMEGA CENTAURI, GREATEST OF GLOBULAR CLUSTERS

A globular cluster is a whirling group of about 100,000 stars occupying a globular-shaped region. To the naked eye Omega Centauri looks like a small, hazy, circular patch. A small telescope confirms its starry structure and a large telescope expands it into an amazing spectacle. Omega Centauri is the most impressive globular cluster in the entire sky.

It is at the end of a line from Beta Centauri to Epsilon Centauri and extended an equal distance beyond Epsilon Centauri (No. 11 in Chart I).

NGC 5138, A FINE CLUSTER FOR BINOCULAR VIEWING

This cluster lies half way between Beta Crucis and Beta Centauri (No. 10 in Chart I).

ANTARES, A DOUBLE STAR

As mentioned in Chapter 5, Antares has two white companion stars. These, however, merely seem to be close to Antares in our line of sight and are not gravitationally linked to Antares. Nevertheless Antares is a double star whose fainter component is visible only in fairly large telescopes. The component looks green but this colour is a visual effect of contrast and does not correspond to its real colour of pale yellow. (No. 13 in Chart IV.)

OTHER FAMOUS OBJECTS IN THE VICINITY OF SCORPIUS

M4, a globular cluster near Antares (No. 12 in Chart IV). (Note: The name M4 means that the object is Number 4 in the Messier Catalogue). Beta Scorpii, an easy double (No. 11 in Chart IV).

M7, a cluster visible to the naked eye (No. 15 in Chart IV).

THE SAGITTARIUS REGION CONTAINING THE GALACTIC CENTRE

When we are looking towards the Sagittarius region bordering on Scorpius (see Chart IV), we are facing the centre of the Milky Way Galaxy where, in our line of sight, the sky seems to be packed tight with a great variety of stars and star patterns. An observer, warmly clad and lying back on a couch, can guide his binoculars to sweep this part of the sky at leisure, finding unlimited entertainment and delight in the unfolding panorama.

THE LAGOON NEBULA, M8

M8 is tucked away in one corner of Sagittarius. To the unaided eye it is a light patch. A telescope enlarges it into an interesting object. It becomes a glorious sight when photographed with a camera attached to one of the world's largest telescopes (No. 16 in Chart IV).

SPRING

September, **October,** November

47 TUCANAE CLUSTER

Another gem of the southern skies is the 47 Tucanae Cluster lying near the Small Magellanic Cloud which is also called Nubecula Minor. 47 Tucanae almost rivals the great Omega Centauri. Both are magnificent globular clusters. (No. 1 in Chart I.)

ate 19. 47 Tucanae. This globular cluster in e Small Magellanic Cloud can be detected thout optical aid as a hazy starlike object, but a large telescope it blossoms out as a owded mass of thousands of stars.
Mount Stromlo and Siding Observatory

Plate 20. Telescopic view of the principal stars in the constellation Musca. Find this constellation also to the right of the Southern Cross in plate 3. The pair of stars at top right corner appear as a single star to the naked eye.

THE DISCOVERY OF VARIABLE STARS

Nearly 400 years ago Pastor David Fabricius, observing the star Omicron Ceti night after night, discovered to his amazement that it was gradually becoming dimmer. After some weeks the star disappeared.

Thirteen years went by and Fabricius spotted it again. He could not explain the star's variations in brightness but he called the star Mira Ceti, the word **mira** meaning wonderful. If he had possessed a telescope he would have noticed that Mira never completely disappears. Mira is an example of a variable star.

Since the time of Fabricius many variable stars have been discovered and carefully studied.

Variable stars are those which vary in brightness either regularly or irregularly. They are classified in two main groups: **extrinsic variables** and **intrinsic variables.** In the first group the variations in brightness are brought about by some process which is external to the star, such as happens when one component of a double star is being periodically eclipsed by its gravitationally linked companion. In the second group the variations are due to changes in the nature of the star, such as pulsations of expansion and contraction. Mira belongs to the group of intrinsic variables.

IMPORTANCE OF STUDYING VARIABLE STARS

Much valuable research work with variable stars is carried out by amateur astronomers. The results are sent to professional organisations where they are studied and interpreted, thus increasing our knowledge of the nature and behaviour of stars.

MIRA OR OMICRON CETI

Mira can be found at the apex of a right-angled triangle whose base is the line joining the Hyades to the Pleiades. The right angle is at the Pleiades and the line from Pleiades to Mira is a little longer than twice the base.

When at its brightest in its 331-day cycle Mira appears as a red star about as bright as the star Peacock (see Chapter 2). At its dimmest it needs a telescope for detection (No. 2 in Chart II).

ALGOL, THE DEMON STAR

This was the first eclipsing variable star discovered and the true explanation for its variations in brightness was supplied by Goodricke in 1783. It is a double star in which one component, swinging round the other, eclipses it at regular intervals so that variations are caused in the brightness of the pair.

Algol, or Beta Persei, has a cycle of 2.9 days and, though faint, it is always within naked-eye visibility. It can be found by envisaging an isosceles triangle with Alderbaran at the apex and with baseline westwards from Capella to Algol. Algol is seen best by observers north of Latitude 35°S. (No. 3 in Chart II.)

THE CRAB NEBULA, M1

The Crab Nebula is very famous indeed. It is composed of the remnants of a supernova which exploded with extreme violence about 5000 B.C. Light from the explosion reached the earth in A.D. 1054 when it was observed by Chinese astronomers as a "new" star which appeared suddenly in the sky.

Within this nebula is a pulsar which is an extraordinary kind of star emitting pulses of light, radio waves and X-rays at the rate of one pulse every 0.0032 seconds.

Through a moderate-sized telescope the Crab Nebula can be seen as a faint oval patch near the star Zeta Tauri which is the unlabelled star just above the figure 8 in Chart II. Long-exposure photographs taken with cameras attached to very large telescopes show the Crab Nebula's explosive appearance caused by shock waves which are still travelling outwards at the rate of 112 million kilometres a day.

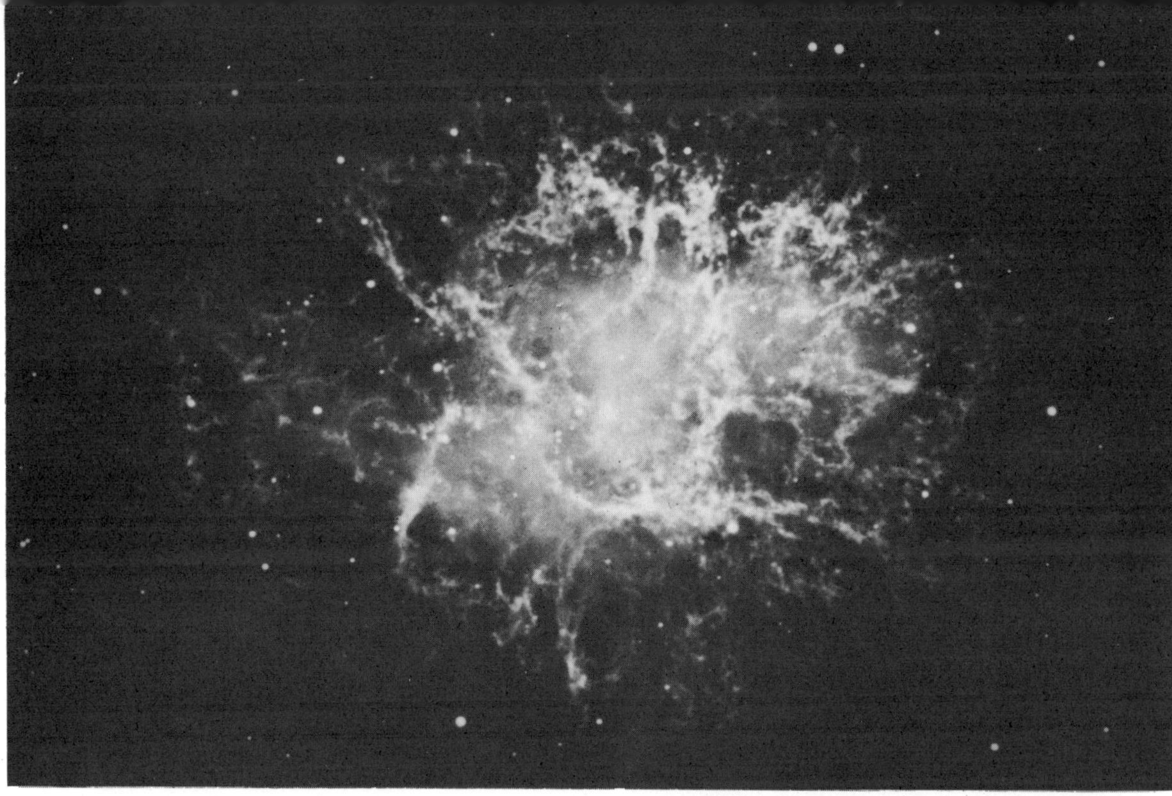

Plate 21. The Crab Nebula, remnant of a supernova (exploding star). Shock waves from the original explosion are still travelling outwards at about 112 million km per day. *Hale Observatories*

THE WHOLE-YEAR CHARTS

The celestial sphere as depicted in Fig. 11 page 8, is a three-dimensional concept which cannot be represented on a flat surface without some distortion, any more than a hollowed-out orange can be pressed on to a flat surface without tearing.

However, special techniques enable cartographers to construct star charts on a flat page in such a manner as to minimise distortion, and in some cases, achieve a gain in the usefulness of such charts. Examples are seen in the Whole-Year Charts.

DECLINATION, RIGHT ASCENSION, VERNAL EQUINOX, ECLIPTIC

Just as in geographic maps the location of a town is indicated by its **latitude** (distance in degrees north or south of the Equator) and **longitude** (distance in degrees east or west of the Greenwich meridian), so in astronomical charts the location of a star is shown by its **declination** (distance in degrees north or south of the Celestial Equator) and **right ascension)** (distance in degrees or equivalent hours east of a certain point – vernal equinox – on the Celestial Equator). The vernal equinox is the point where the sun, travelling along its path **(ecliptic)** on the celestial sphere, crosses the Celestial Equator from south to north.

In astronomical charts north declination is plus (+) and south declination is minus (−). Below the charts are lists of important celestial objects with their locations in declination and right ascension.

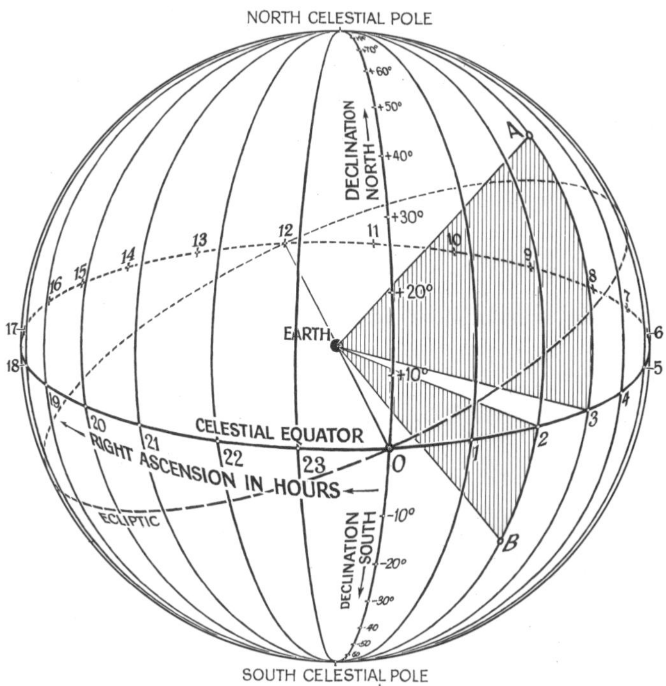

Fig. 26. Location of stars on the celestial sphere. Coordinates of Star A: RA 3h, Dec +40°. Coordinates of star B: RA 2h, Dec −20°. The vernal equinox is the point where the ecliptic cuts the celestial equator at 0 hours right ascension.

WHOLE-YEAR CHART I: SOUTH POLAR REGIONS

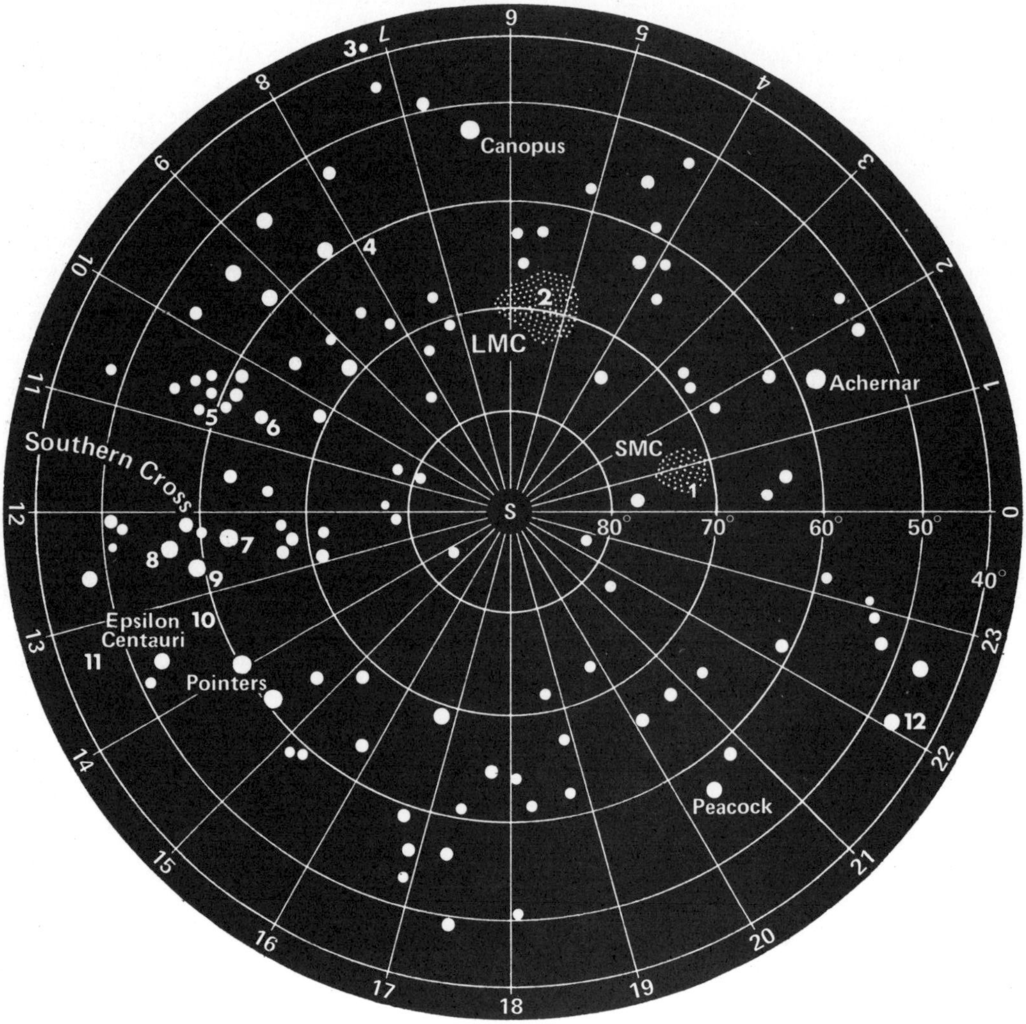

REFERENCE TO NUMBERS ON CHART

Name	Right Ascension hours (h)	Right Ascension minutes (′)	Declination degrees (°)	Declination minutes (′)
1. 47 Tucanae				
2. NGC 2070				
3. L² Puppis	7	12.0	-34	44
4. NG 2516	7	57.5	-60	44
5. Eta Carinae	10	43.1	-59	25
6. Theta Carinae	10	11.0	-64	39
7. Alpha Crucis	12	23.7	-62	49
8. Gamma Crucis	12	28.4	-56	50
9. Kappa Crucis	12	50.7	-60	05
10. NGC 5138	13	24	-58	30
11. Omega Centauri	13	23.7	-47	13
12. Pi Gruis	22	19	-40	37

WHOLE-YEAR CHART II: EQUATORIAL REGIONS

Right Ascension 0 hours to 8 hours

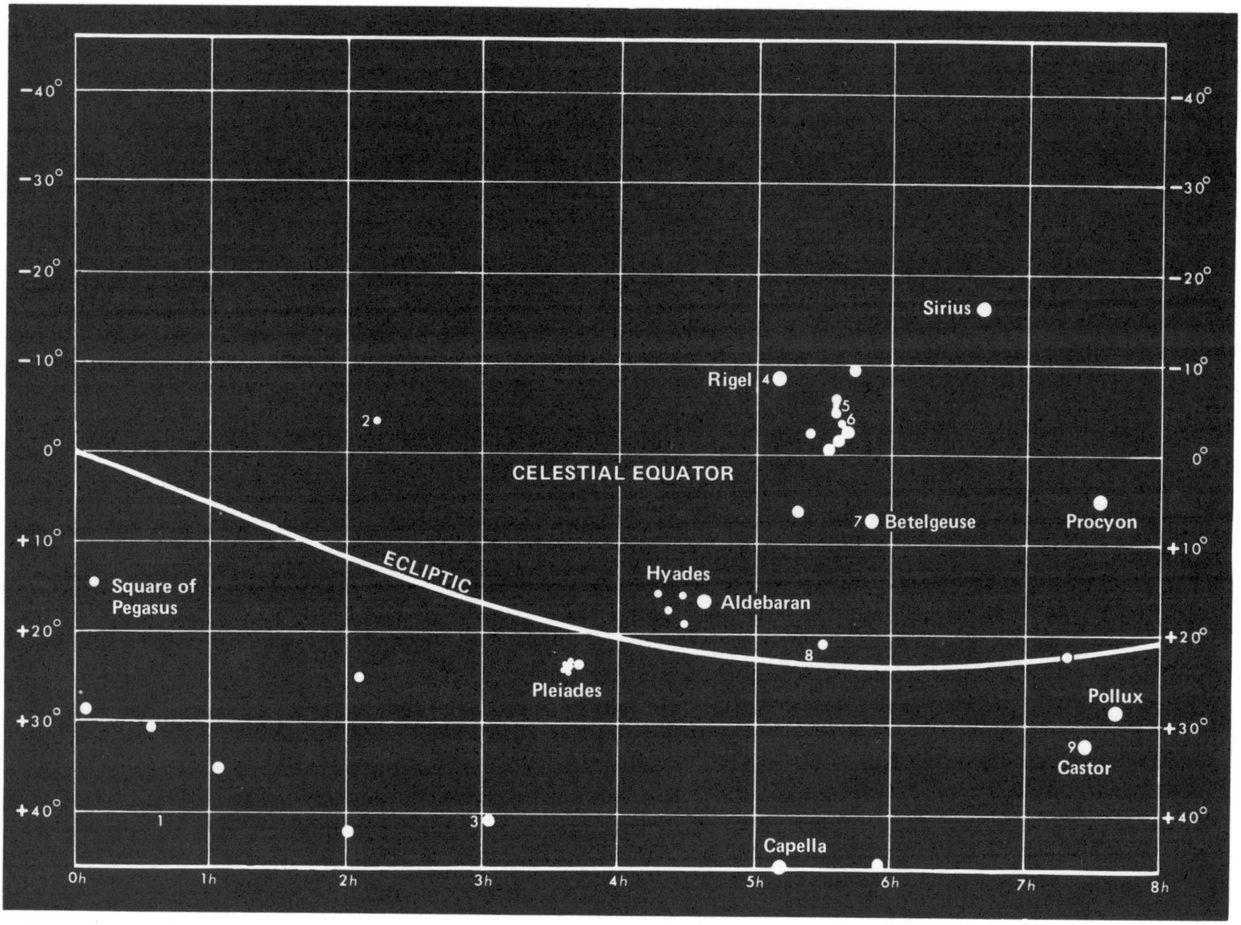

REFERENCE TO NUMBERS ON CHART

Name	Right Ascension hours (h)	minutes (′)	Declination degrees (°)	minutes (′)
1. Andromeda galaxy	0	40.0	+41	00
2. Mira	2	16.8	−3	12
3. Algol	3	4.9	+40	46
4. Rigel	5	12.1	−8	15
5. Gt Neb in Orion	5	32.8	−5	25
6. Sigma Orionis	5	36.2	−2	36
7. Betelgeuse	5	52.0	+7	24
8. Crab Neb	5	31.5	+21	59
9. Castor	7	31.4	+32	00

WHOLE-YEAR CHART III: EQUATORIAL REGIONS

Right Ascension 8 hours to 16 hours

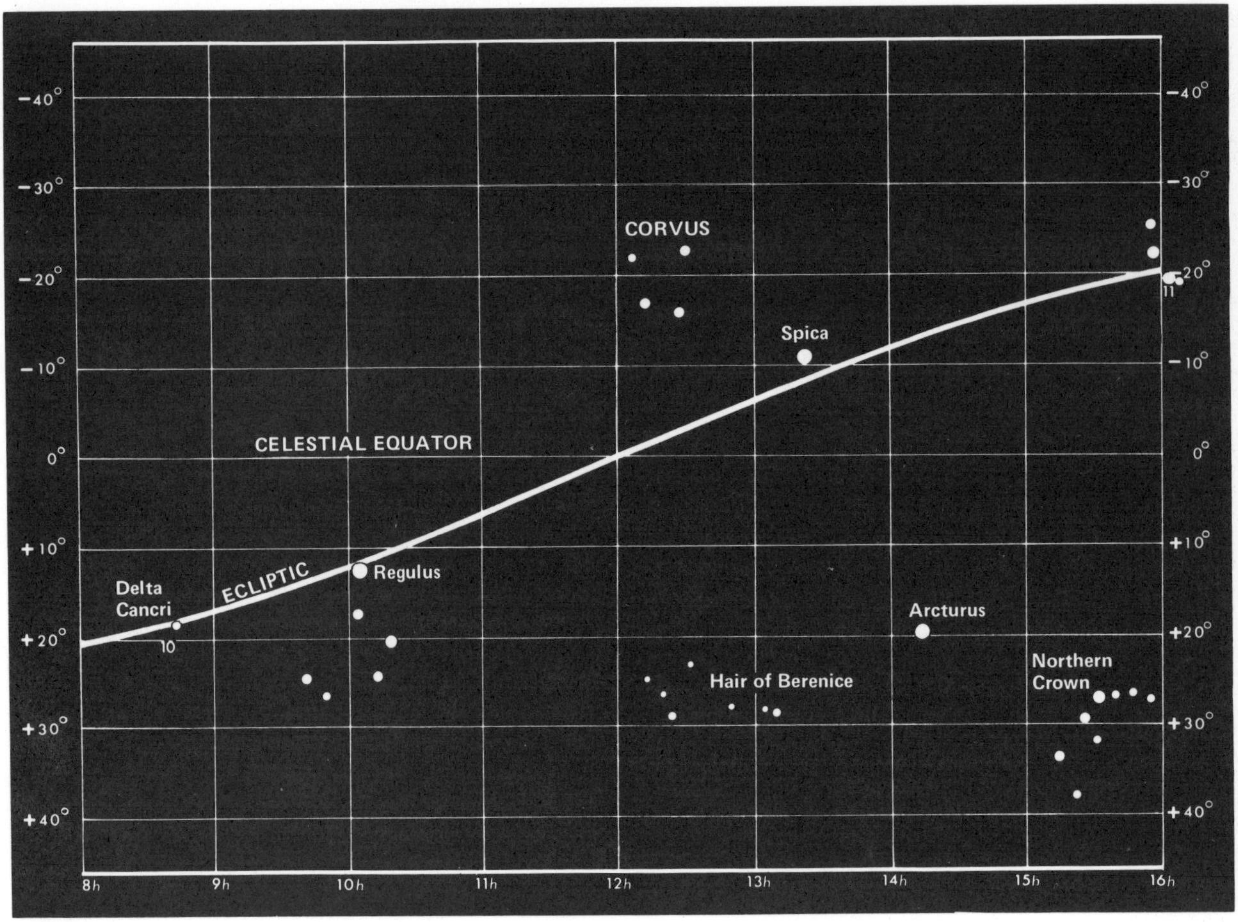

REFERENCE TO NUMBERS ON CHART

Name	Right Ascension hours (h)	Right Ascension minutes (′)	Declination degrees (°)	Declination minutes (′)
10. Praesepe	8	37	+20	10
11. Beta Scorpii	16	2.5	−19	40

WHOLE-YEAR CHART IV: EQUATORIAL REGIONS

Right Ascension 16 hours to 24 hours

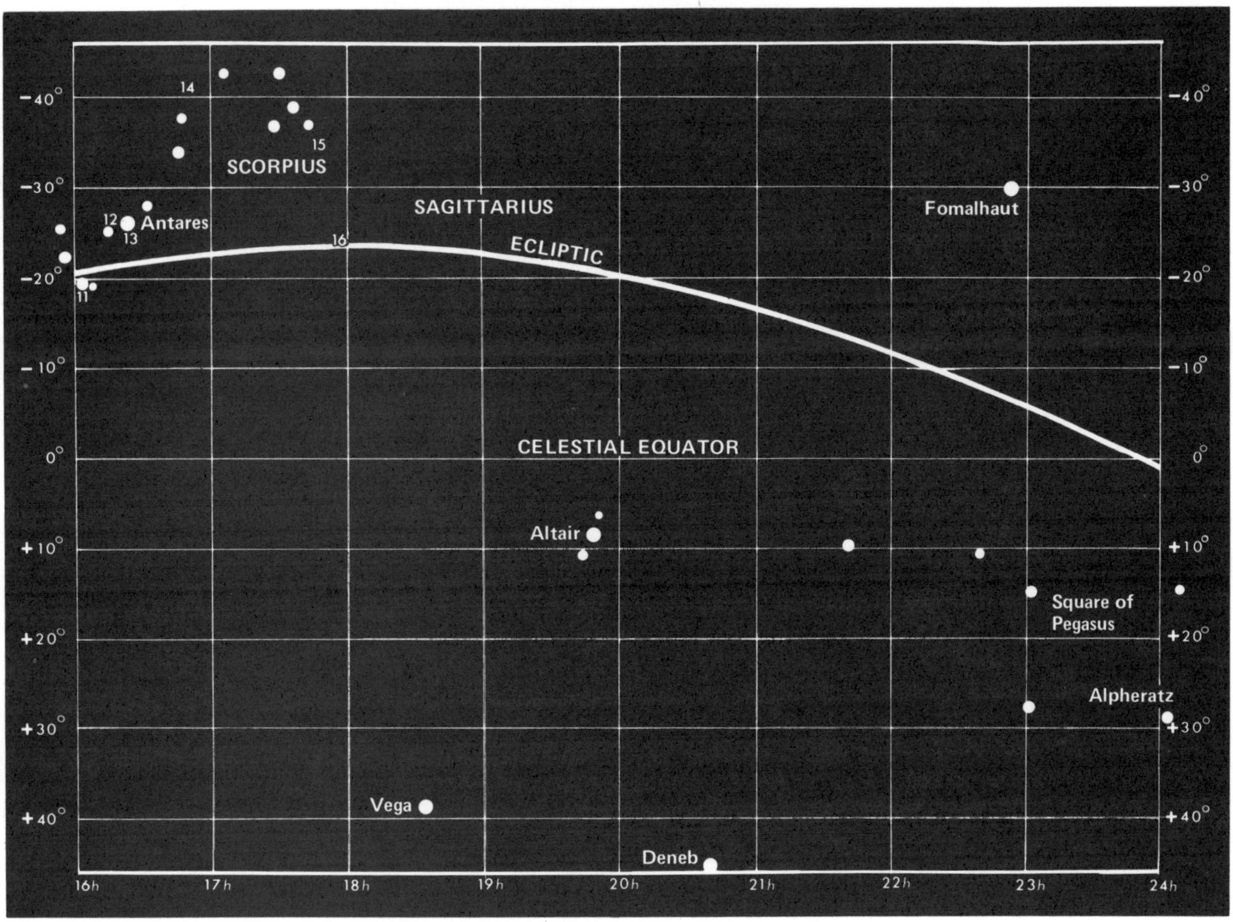

REFERENCE TO NUMBERS ON CHART

Name	Right Ascension hours (h)	Right Ascension minutes (′)	Declination degrees (°)	Declination minutes (′)
11. Beta Scorpii	16	2.5	−19	40
12. M4	16	20.5	−26	24
13. Antares	16	26.5	−26	20
14. NG 6231	16	50.5	−41	43
15. M7	17	50.7	−34	48
16. M8	18	0.1	−24	23

WHOLE-YEAR CHART V: NORTH POLAR REGIONS

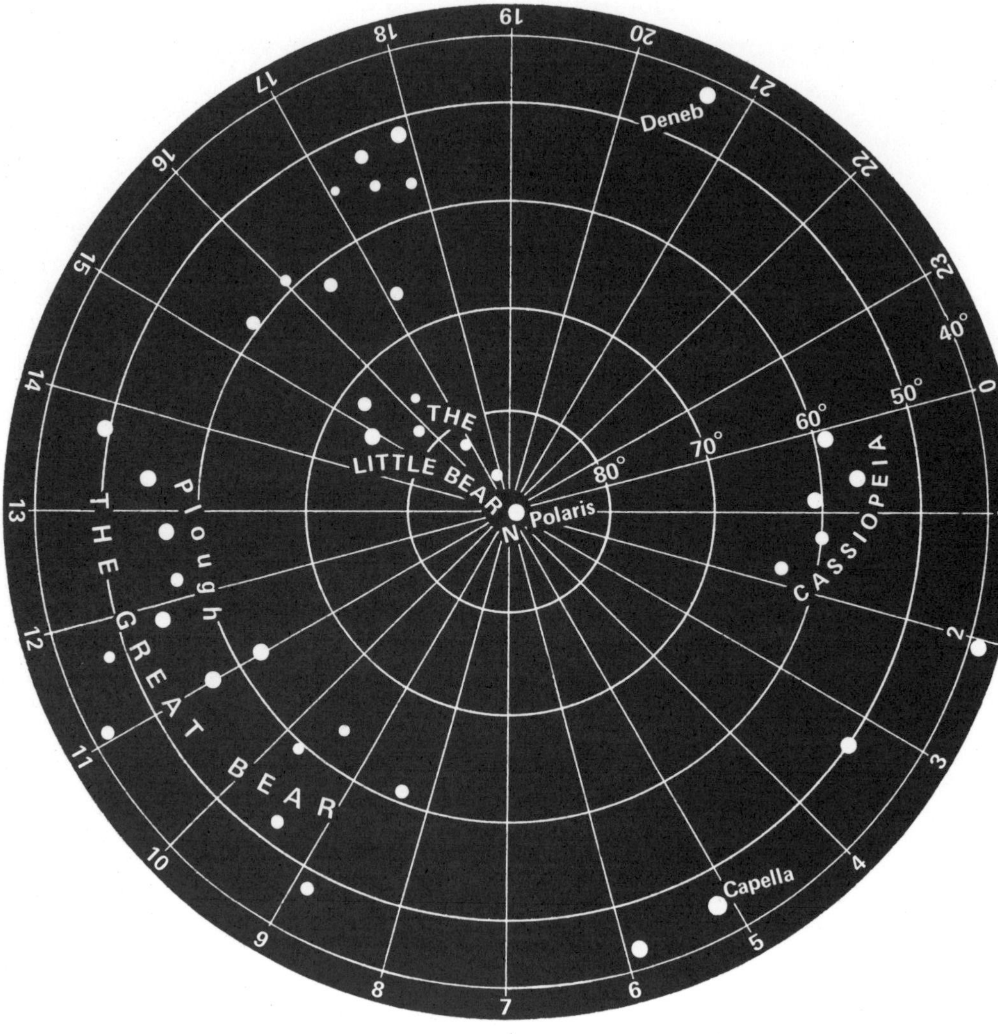

Among the best-known features in the North Polar region are the constellations Ursa Major (the Great Bear), Ursa Minor (the Little Bear) and Cassiopeia (the Sea Goat). These are shown on this chart as well as the Plough, which is a portion of the Great Bear, and the three important stars Polaris (the Pole Star, only 1° from the North Celestial Pole), Deneb and Capella.

VII. THE SOLAR SYSTEM

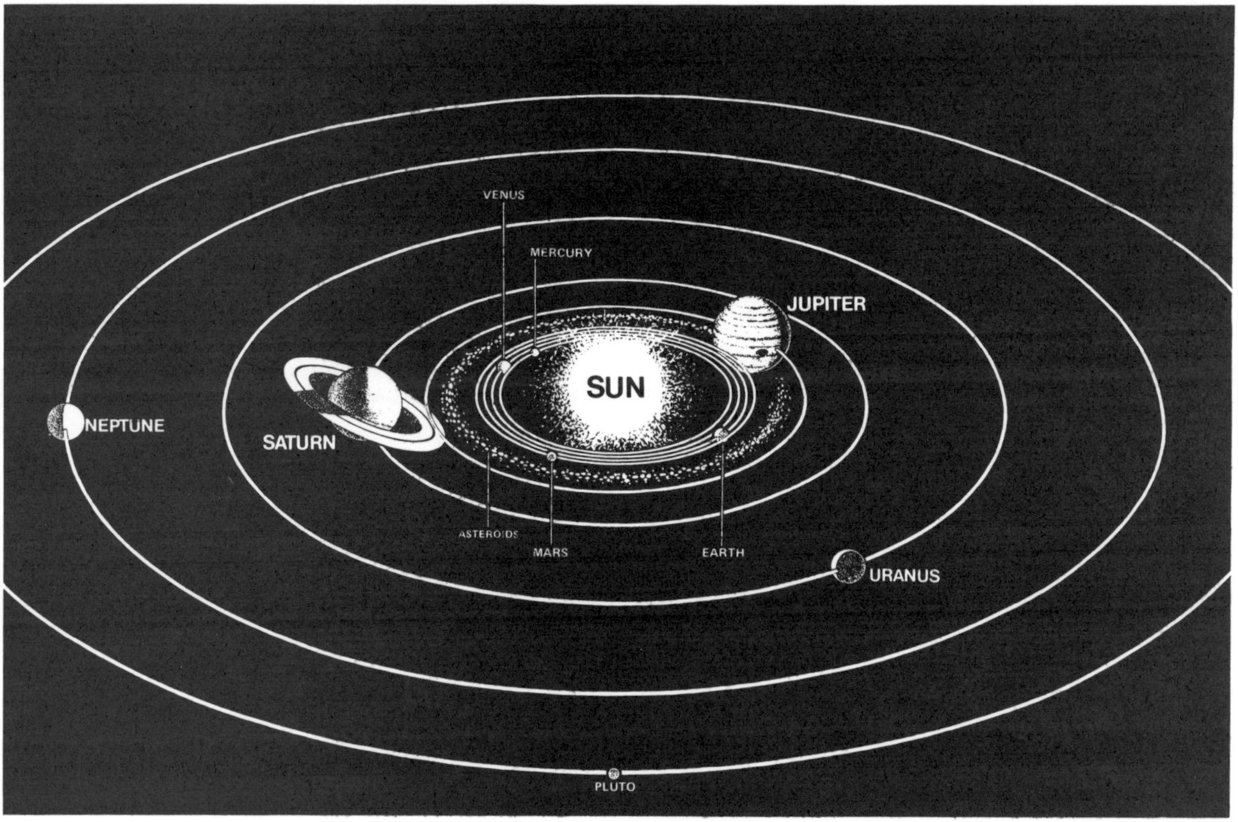

Fig. 27. The Solar System.

MEAN DISTANCES OF PLANETS FROM SUN

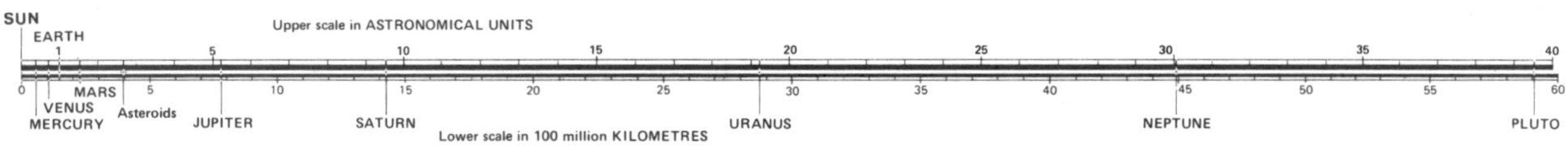

Chapter 8.

AN OVERALL VIEW

If by some magic we could place ourselves far away from the earth so as to see the whole solar system in one grand vista we would realise that it consists mainly of empty space.

The sun would look like a star, which it is, and the planets would appear as dots of light separated by vast distances.

Given ultra-keen eyesight we would discover altogether nine planets, 50 moons, many thousands of minor planets or asteroids, innumerable chunks of rock which are meteoroids, hundreds of comets and great quantities of thinly spread interplanetary dust.

We would notice that the sun maintains gravitational discipline over every object in its domain even though that domain extends outwards as far as, and beyond, the lonely planet Pluto which is at a mean distance from the sun of nearly 6000 million kilometres.

But let us come back to earth and give thought to a number of facts which relate to the orbital movements of the planets and to their other general features. These facts are basic to any theories of the origin of the solar system.

1. The planetary orbits lie approximately in one plane. Exceptions are the orbit of Mercury making an angle of 7° to the plane, and that of Pluto with a 17° angle.
2. The planets revolve round the sun in the same

direction – anticlockwise as seen looking down on the sun's north pole.

3. With two exceptions, Venus and Uranus, the planets rotate round their axes in that same anticlockwise direction.
4. With some exceptions, the moons revolve round their parent planets in that same anticlockwise direction.
5. Our moon rotates on its axis in that anticlockwise direction. Most of the other moons may be rotating in a similar direction.
6. The nearer a planet is to the sun, the greater its orbital speed.
7. As Johannes Kepler discovered in 1604, planetary orbits are ellipses, not circles.
8. The inner planets Mercury and Venus are too near the sun and therefore too hot to support life as we know it. The outer planets, Mars, Jupiter, Saturn, Uranus, Neptune and Pluto are too far from the sun and therefore too cold to support life as we know it.
9. The planets Mercury, Venus, Earth and Mars consist chiefly of rock. Jupiter, Saturn, Uranus and Neptune are made principally of gas or of gaseous material which is in a frozen state. The composition of the outermost planet Pluto is not known.

Note: See Appendix F for details of planetary sizes, distances and other items of information.

ORIGIN OF THE SOLAR SYSTEM

Let us consider one widely held theory, the nebular hypothesis. We can imagine a situation in which a vast cloud of dust and gas is floating in space. It could be material blasted out from a supernova or exploding star. Gradually the particles in the cloud are brought closer together by gravity. The whole mass takes the shape of a disk rotating at increasing speed as it contracts. Whirlpools of the material become separate entities as they condense to form proto-planets, or planets in the making. Nearby, lesser whirlpools form proto-moons. Eventually these objects develop into the planets and moons of the present solar system.

In the meantime the relatively great abundance of material in the central regions of the cloud acquires a high temperature as its atoms jostle each other with increasing energy. After many millions of years this proto-sun, consisting chiefly of hydrogen, becomes so compressed that the temperature rises to the point where hydrogen nuclei begin to fuse in thermonuclear reactions. At that point the central mass has become the fiercely burning celestial furnace that we call a star or sun.

WHAT HOLDS THE SOLAR SYSTEM TOGETHER?

While the solar system was being formed, every planet acquired its own particular speed in orbit. In the case of the earth, the present orbital speed is nearly 30 km per second. If the gravitational attraction between the sun and the earth were switched off (not possible, of course!), the Earth would leave its orbit and race away in a straight line at nearly 30 km per second. However, gravity in fact exerts a steady pull towards the centre of the sun. This balances the earth's tendency to travel in a straight line and keeps the earth in its present orbit. The situation resembles that in which a person swings a stone attached to the end of a string. The stone has a tendency to travel in a straight line but the pull exerted by the person on the string changes that straight line to a circular path.

THE ORRERY

An orrery is a clockwise model of the Solar System. It is named after Charles Boyle, fourth Earl of Orrery, patron of John Rowley who constructed the original instrument in 1713.

The orrery shown in Plate 22 can be used to demonstrate several features of the solar system. For example, when the controlling handle is turned the planets travel at their correct relative speeds round the sun. As another example, it will demonstrate that Mars makes a close approach to the earth every 2 years 2 months approximately.

An orrery of much simpler construction can be made by placing "planets" at the end of different lengths of wire bent at right angles and supported on a central pillar. The "planets" in such a model must be moved separately by hand.

Elaborate orreries are not now manufactured for sale. They have been superseded by planetarium projectors which are very costly but very efficient.

Plate 22. The author displaying an antique orrery (clockwork model of the solar system).

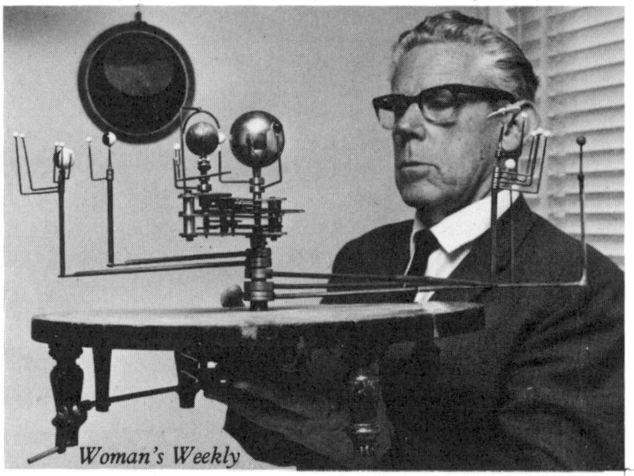

OTHER SOLAR SYSTEMS?

It is quite likely that many other solar systems exist throughout the universe. Some may contain planets supporting a variety of life forms.

Chapter 9.

THE SUN – SOURCE OF LIFEGIVING POWER

Although the sun is nearly 150 million kilometres away from us we dare not risk our eyesight by looking directly at it. What sort of an object is the sun that it can shine with such devastating brilliance?

MYSTERY OF THE SUN'S INTERIOR

The principal mechanism for energy production in the interior of the sun is the proton-proton chain in which protons (hydrogen nuclei) fuse to produce helium nuclei. In the process some of the material is annihilated and transformed into energy such as light and heat. This transformation takes place in accordance with Albert Einstein's famous formula $E = mc^2$ which tells us how much energy (E) is produced when a mass (m) of matter is destroyed. c^2 is the square of the velocity of light. In the case of the sun a colossal amount of material, four million tonnes, is annihilated every second. The sun's interior is maintained at a temperature of 15 to 20 million degrees.

Fortunately for us the sun, having a mass of 2,000 million million million million tonnes, can afford such prodigality. At its present rate of destruction it is being reduced by only one-tenth of one per cent of its mass in 15,000 million years.

There is a mystery associated with the proton-proton chain. A "ghost particle" should, according to theory, be emitted in the process. It is the neutrino which, almost unbelievably, has no mass when at rest and no electric charge. Yet it carries energy and it could pass through 1000 light-years of solid lead. Some neutrinos should be raining down on the earth but, so far, astronomers have not been able to detect them.

PHOTOSPHERE, CHROMOSPHERE, PROMINENCES

The sun, being entirely gaseous, has no solid surface but the part we see is the **photosphere** or "light" layer of its atmosphere.

Above the photosphere of the sun is the **chromosphere** which is invisible except when the sun's bright disk is blotted out during the few minutes of a total eclipse of the sun. In that short period

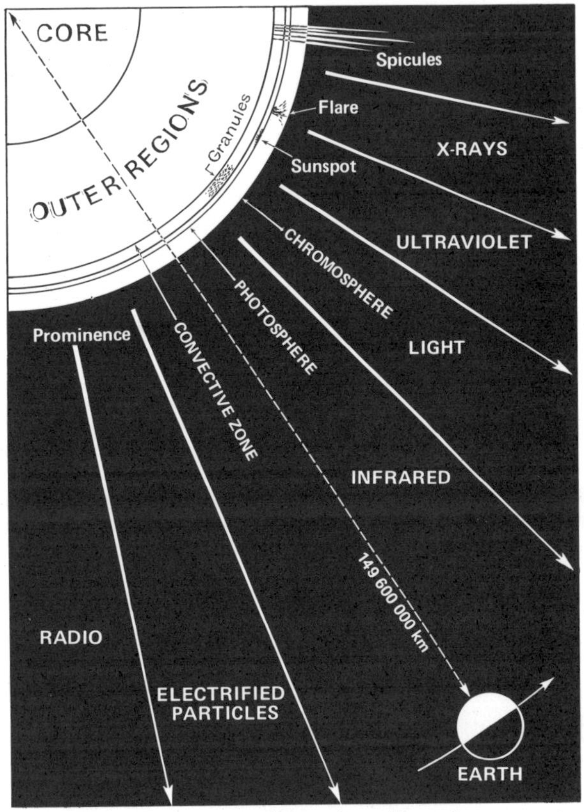

Fig. 28. The sun continuously emits vast quantities of energy in the form of electromagnetic radiations as well as electrified particles which constitute the solar wind.

the chromosphere is revealed as a glowing red circle encompassing the black disk of the eclipsing moon. The red colour indicates the present of hydrogen.

Time-lapse movie photographs of a total solar eclipse sometimes record **prominences** bursting forth from the chromosphere and attaining heights above the sun of hundreds of thousands of kilometres. Prominences can take the forms of sheets, sprays, tongues or jets of flame shooting almost straight upwards or curving in noble arcs before falling back into oceans of fire. See Plate 23.

Plate 23. A solar prominence. This explosion, one of the largest ever recorded, was photographed in June 1946. Within about two hours it streamed to a height of over a million km above the sun's surface.

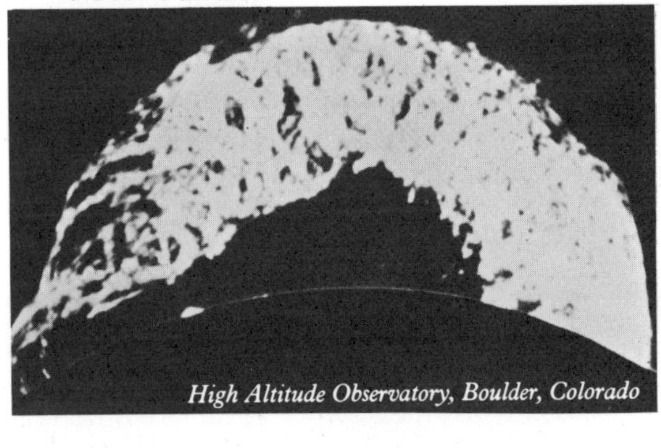

High Altitude Observatory, Boulder, Colorado

CORONA

Above the chromosphere is the sun's **corona** which, during a total solar eclipse, is seen as a breath-taking, pearly-white mass much wider than the chromosphere. It consists of electrically charged particles jostling each other at a temperature of about one million degrees. Tenuous and invisible parts of the corona extend beyond the earth. See Plate 24, Plate 25.

Plate 24. The totally eclipsed sun, its white corona surrounding the moon's black disk. Jupiter can be seen at top right of eclipsed sun.

. O. Williams

Plate 25. The Diamond Ring effect.
H. O. Williams

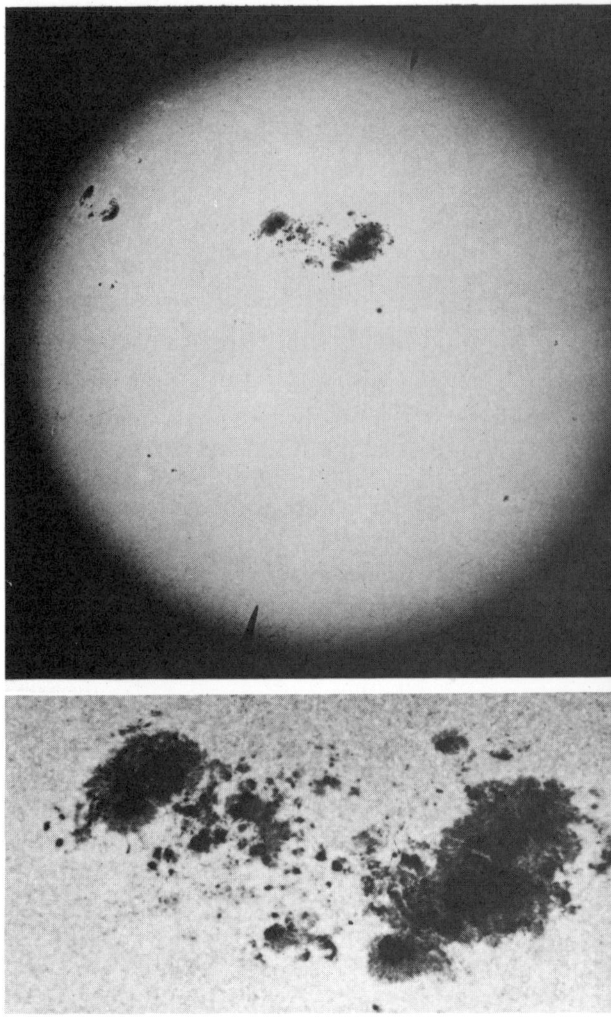

Plate 26. An exceptionally large sunspot recorded on 7 April 1947. In the lower frame is an enlarged view of the same sunspot.

SUNSPOTS

Several centuries ago Chinese astronomers peering through fog discovered on the sun some impermanent areas which we now call **sunspots**. They are really very bright areas but they appear dark only when the sun's disk has been dimmed by fog or when special viewing methods such as projecting the sun's image on to a white screen are used. The temperature of a sunspot is about 1000° less than that of the photosphere.

Sunspots are thought to be whirlpools of flaming, electrified particles stirred into fierce activity by the interplay of electrical, magnetic and gravitational forces.

Patient observation of sunspots over a period of several days, using a projection method as shown in Plate 27, can easily be carried out by an amateur astronomer. The movements of the sunspots indicate that the sun is rotating.

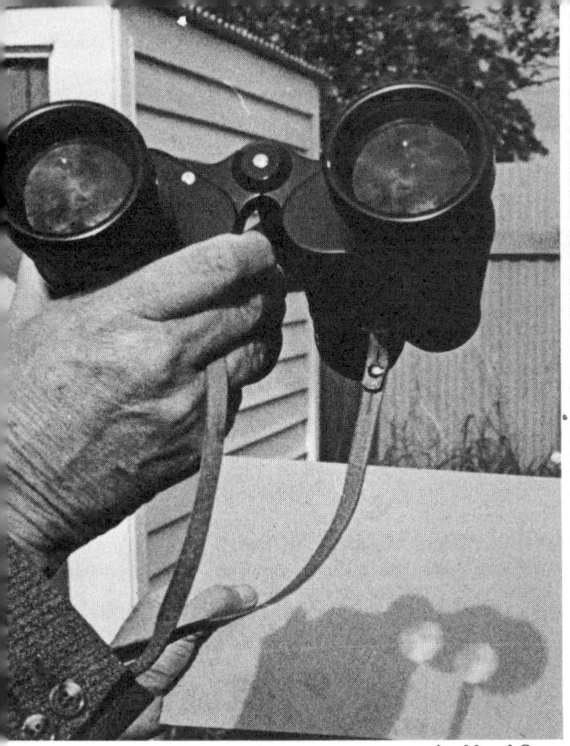
Auckland Star

A. L. Rounthwai

Plate 27. The safe projection method of examining the sun's image for sunspots or for the progress of a solar eclipse, using (left) binoculars or (right) a telescope.

FLARES

Now and again parts of the photosphere are stirred into intense activity to form a **flare**, often associated with large sunspots. From the flare are blasted enormous quantities of electrified particles and electromagnetic waves which could be hazardous to astronauts caught without adequate protection. Flares can cause disturbance to the earth's magnetic field and can be responsible for radio fade-outs. Occasionally flares give rise to the beautiful and awesome phenomena of the Aurora Borealis ("northern lights") or Aurora Australis ("southern lights").

RADIATIONS FROM THE SUN

Emitted from the sun are rays representing all parts of the electromagnetic spectrum: gamma rays, X-rays, ultraviolet rays, light rays, infrared rays and radio rays. These all travel at the speed of light. Light, for example, takes just over eight minutes to reach the earth. Thus, if the sun's light were suddenly switched off we would know nothing about it until over eight minutes later. See Fig. 29. The sun also blasts outwards in all directions electrically charged particles including protons and electrons which constitute what is called the **solar wind.** It streams across vast sections of the solar system.

Chapter 10.

THE EARTH – PARADISE OF THE SOLAR SYSTEM

Our home in the universe is a ball of rock travelling at great speed on an endless journey round the sun. It carries as passengers more than 4000 million human beings as well as other living organisms too numerous to count.

GRAVITY

With a tremendous mass of 6000 million million million tonnes the earth possesses sufficient gravity to hold itself together and to prevent all movable objects such as ourselves from drifting away into space. In order to escape the earth's gravitational field with one burst from its rockets a spaceship would need a take-off speed of 11.8 km per second. That is the velocity of escape from the earth, calculated without reference to the presence of the atmosphere.

Fig. 29. The Earth in orbit round the sun.

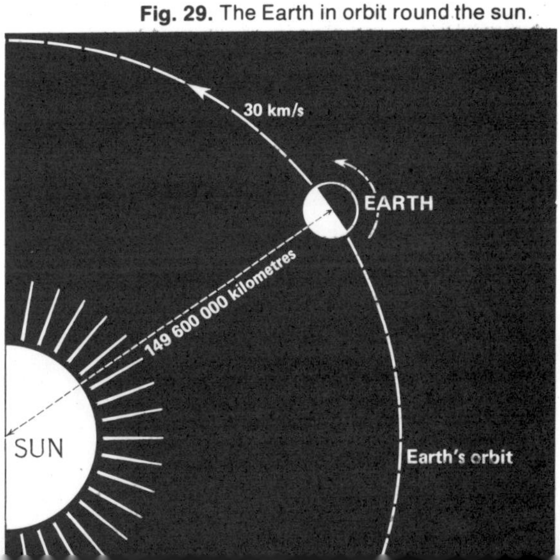

TEMPERATURE

The earth receives just the right amount of warmth from the sun to maintain life in all its forms. But scientists have estimated that if the average daily temperature of the surface of the earth were to rise or fall by as little as 20°C the overall ecological balance would be upset to the extent of bringing all life here to an end.

DISTANCE FROM THE SUN

One reason why the average temperature is favourable is that we are at a suitable average distance from the sun – nearly 150 million kilometres. If the Earth wandered too far away from the sun, or too close to it, the changes in temperature would be disastrous. As things are, the earth's elliptical orbit never brings it nearer to the sun than 147 million kilometres, nor further away than 152 million kilometres.

SPEED IN ORBIT

To maintain the present average distance from the sun we must travel at a suitable average speed in orbit which is 30 km per second. If an aircraft could travel at that enormous speed it could take us from Auckland to Sydney and back in about two minutes.

It will be realised that the earth's temperature, its distance from the sun and its orbital speed represent interlocking factors which, within relatively narrow limits, are perfectly balanced to provided us with the right environment for the maintenance of life.

THE EARTH'S SPIN

The earth rotates at the rate of one complete turn every day. If this spin were to slow down appreciably, the overall climatic conditions would be dangerously altered. If the spin were to speed up too much the increased centrifugal forces would throw people and other unattached objects from the earth's surface like particles of soil flung off the tyre of a spinning bicycle wheel.

ATMOSPHERE

The gases now present in the atmosphere, chiefly nitrogen and oxygen, are necessary for our survival. At a high level a layer of ozone protects us from most of the dangerous ultraviolet rays emanating from the sun. The atmosphere is, however, transparent to light which permits photosynthesis to occur in plants. That transparency enables us to see parts of the universe beyond the earth.

All in all, the earth in its unpolluted state is a home with countless opportunities for enjoyable living. Unlike the other planets with their hostile environments the earth is unquestionably the paradise of the solar system.

Chapter 11.

THE MOON OUR FUTURE HOME?

We've all looked at the moon many times and have occasionally paused to enjoy its serene beauty. But comparatively few people have explored it through a telescope which can transform its almost featureless surface into a glowing scene of plains, mountain ranges, craters, pits, rills and other intriguing features.

Newcomers confronted with such a view for the first time tend to express their surprise and delight. Then, given a photograph of the moon as in Plate 28, and a few minutes at the telescope, they can spot the principal lunar features and become acquainted with them by name. It is important to realise that our southern hemisphere view of the moon is upside down compared with the northern hemisphere view. The reason for this will be obvious from Fig.15.

SURFACE FEATURES

The relatively dark areas are maria or seas, so called by early moon gazers who guessed wrongly in their efforts to explain these features. Maria are comparatively flat plains of lava.

SUITABLE TIMES TO OBSERVE THE MOON

The features on the lunar surface stand out most clearly when light rays from the sun are striking the surface at angles which ensure that long shadows of mountains and crater walls are being cast on the surface. This is the situation at any time when the moon is visible except at, or close to, full moon. Nevertheless the full moon is best for viewing the crater Tycho with the streaks of splash material extending radially from it. See Plate 29 (d). For full descriptions of lunar features consult a suitable book on general astronomy, remembering that if the book has been published in the northern hemisphere its moon charts are likely to be upside down from our point of view.

PHASES OF THE MOON

If we stand out of doors on a sunny day and, with our back to the sun hold out a tennis ball at arm's length, we will notice that the illuminated half of the ball faces us. The ball, representing the moon, will be at the phase of "full moon". If we now turn slowly anticlockwise for a quarter circle, still holding the ball at arm's length, we will see that the ball is at the phase "first quarter". If we had turned clockwise for a quarter circle the ball would have shown "last quarter".

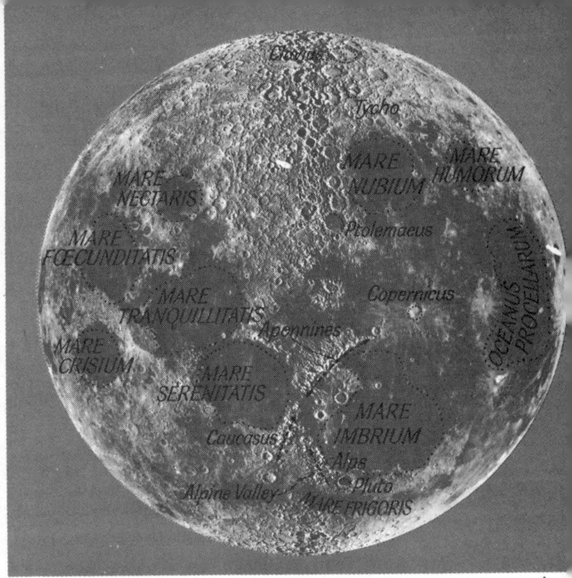

←A **Plate 28. A & B.** The moon. A composite photograph of the moon's left half taken near first quarter and of the right half taken near last quarter. The shadows thus created by the sun's slanting rays give much clearer detail than is possible with just one photograph of the flat looking full moon. *Photos: Lick Observatory* ↑B

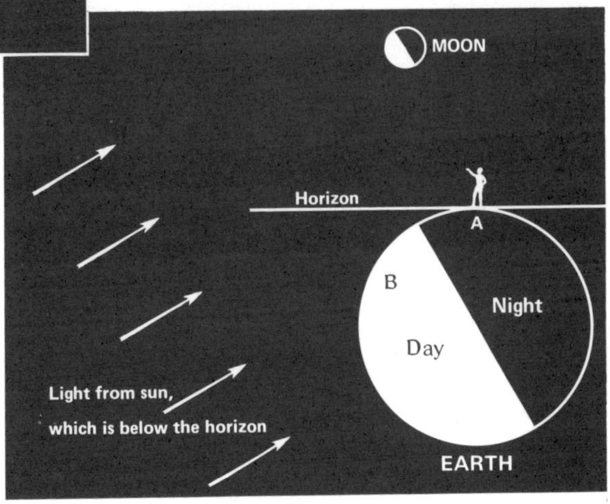

Fig. 30. Viewing the moon. At night the moon is illuminated by the sun, which is below the horizon. An observer at A sees the moon at first quarter, that is he sees a disk of which the left half (quarter of the whole moon) is illuminated. If, in this diagram, the sun were almost directly below the Earth the observer would see a full moon with the disk completely illuminated. The moon is frequently visible in daylight from such a position as B in the diagram.

PRINCIPAL MARIA:

1. Mare Nectaris (Sea of Nectar)
2. Mare Foecunditatis (Sea of Fertility)
3. Mare Crisium (Sea of Conflicts)
4. Mare Tranquillitatis (Sea of Peace)
5. Mare Serenitatis (Sea of Serenity)
6. Mare Frigoris (Sea of Cold)
7. Mare Imbrium (Sea of Rains)
8. Oceanus Procellarum (Ocean of Storms
9. Mare Humorum (Sea of Moisture)
10. Mare Nubium (Sea of Clouds)

PRINCIPAL CRATERS:

1. Clavius (German mathematician)
2. Copernicus (Polish astronomer)
3. Plato (Greek philospher)
4. Tycho (Danish astronomer)

PRINCIPAL MOUNTAINS:

1. Apennines
2. Caucasus
3. Alps

Compare these phases, and intermediate phases, with those of the real moon as indicated in Plates 29(a) to (g). The term "new moon" in the astronomical sense refers to the situation when the moon is presenting its shaded half to us, that is, when it is invisible.

Because the moon revolves round the earth in the same period as it rotates on its axis it always presents the same face to the earth – except for the moon's libration which is a slight nodding (up and down) and a swinging (from side to side).

CONDITIONS ON THE MOON

Because there is virtually no atmosphere on the moon it is a world with no life, no clouds, no rain, no coloured sunsets or sunrises, no blue sky (the sky is black and the stars are visible even in daytime), no protection from dangerous rays coming in from the sun, no sound.

If an observer on the moon could have watched, from a safe distance, the 15-tonne 3rd-stage component of the **Apollo 15** spacecraft crashing on to the moon at 8000 km per hour as part of a moonquake experiment, he would have been awed by the sight but no sound would have reached him. Astronauts must use portable radio equipment in order to communicate.

Plate 30. Portion of the Lunar Alps, separating Mare Imbrium (Sea of Rains) from Mare Frigoris (Sea of Cold). The walled plain Plato is about 96 km across. The Alpine Valley, 133 km long and from 5 to 10 km wide, looks like a gash across the Lunar Alps. *Lick Observatory*

COLONISING THE MOON

For a long stay of months or years on the moon elaborate buildings would be required for housing, and factories would be necessary for air and food. Preferably the buildings should be underground.

There is no free water on the moon and no trace of moisture in the rocky surface. However, water could be manufactured by causing oxygen from the rocks to combine with hydrogen transported from the earth. To obtain free oxygen for other purposes, such as breathing, the water already produced could be passed through electrolytic cells which separate the oxygen of the water from the hydrogen. In this process the hydrogen could be used over and over again. Air would be produced by mixing nitrogen (obtainable from lunar rocks) with oxygen already manufactured. Electricity would come from generators powered by nitrogen engines instead of steam engines.

Some astronomers think that human habitations will be set up on the moon within the next decade or so and this could lead to wide-scale colonisation there. If pollution or nuclear war drives man from the earth his only chance for survival might well be to build a new home on the moon.

Chapter 12.

CLOSE LOOK AT THE PLANETS

In recent years the study of planets has been transformed by new techniques in optical and radio astronomy, and by the use of camera-carrying spacecraft to take a close look at several alien worlds.

Photographs of the surfaces of Mercury, Venus and Mars have removed some of the former glamour from these planets but they have brought deep satisfaction to scientists who prefer facts to speculations. Records from instruments other than cameras have enabled astronomers to analyse the composition of planetary atmospheres, to work out the climatic conditions and to assess the possibilities of the existence of living organisms.

In this chapter the nine planets are presented in the order of their increasing distances from the sun, starting with the nearest which is Mercury. In each case a few distinctive features have been selected. Additional information is provided in the tables incorporated in Appendix F.

MERCURY

In Greek mythology Mercury was the fleet messenger of the gods; his name was appropriately given to this planet which is the fastest moving of all

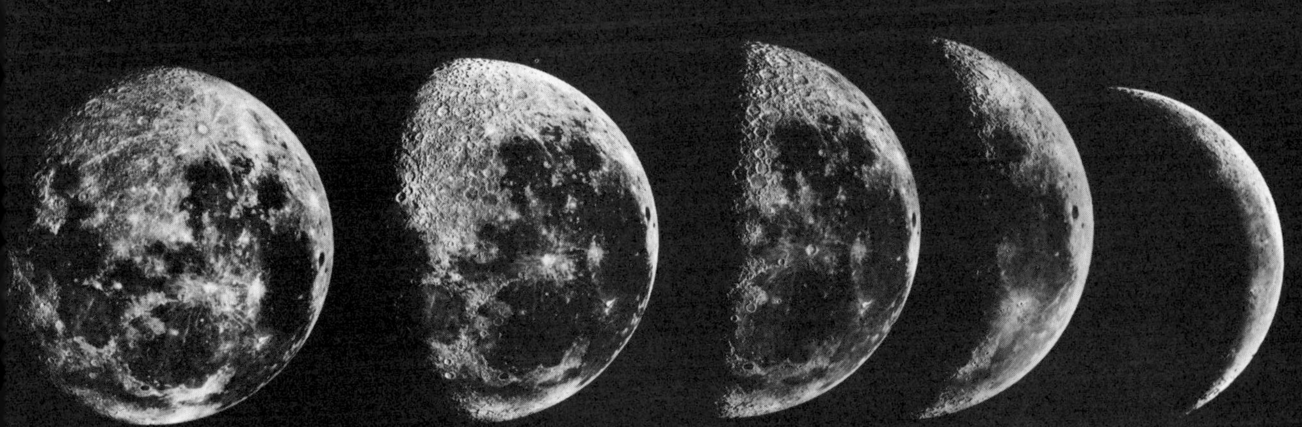

Age 17·8 days — Age 20.01 days — Age 22 days (last quarter) — Age 24 days — Age 26 days

the planets. Mercury, much smaller than the Earth, has always been regarded as an elusive object. Relatively few people have seen it though it is easy to find if you know when and where to look for it – as explained in Chapter 13.

All that can be seen of Mercury through a large telescope are a few vague patches on its surface, but no definite markings which would enable astronomers to discover the rate at which it turns on its axis. However, from those patches an assumption was made that Mercury rotated at such a rate that it always presented the same face to the sun. If that were true the planet's rotation period would be equal to the period taken to revolve round the sun, namely 88 days.

That idea had to be discarded when, in 1965, radioastronomers used radar to pinpoint definite topographical features rotating with the planet. With this technique Mercury's rotation period was found to be 58.67 days, not 88 days.

The year 1974 brought an exciting breakthrough in man's exploration of Mercury. The American spacecraft **Mariner 10** passed within a few hundred kilometres of the planet and sent back to earth thousands of clear pictures showing a cratered surface resembling that of the moon. Obviously Mercury has suffered severe bombardment from meteoroids during some far-distant period of its history.

VENUS

Venus, named after the goddess of love and beauty, can often be seen shining brilliantly in the evening sky or the morning sky.

It is about the same size as the earth and its orbit is relatively close to ours. Hence it is sometimes referred to as earth's "twin".

This planet is covered permanently with clouds which hide its surface from even the most powerful telescopes. The clouds consist mainly of carbon dioxide with small amounts of water, ammonia and other compounds. Not being able to see the solid surface, astronomers did not have the slightest idea of the planet's period of rotation until radar experiments during 1961-64 solved the problem. Venus rotates once every 243 days but, surprisingly, the direction of rotation is backwards compared with that of other planets. Astronomers call it retrograde rotation. Remembering that the earth's rotation period is approximately 24 hours, we must regard the Venusian period of 243 days as representing a very slow rate of turning.

Before visiting Mercury in 1974, **Mariner 10** flew by Venus and obtained informative photographs of the clouds as well as other useful information. Then, in 1976, two Russian spacecraft **Venera 9** and **Venera 10** put landers on the Venusian surface. **Venera 9's** lander survived the extremes of heat and pressure for 53 minutes; **Venera 10's** lander lasted 65 minutes. Both landers transmitted photographs and additional reports to their respective orbiters which relayed these to earth.

The photographs revealed rock-strewn scenes indicating that geological weathering is active. Relevant instruments recorded very high surface temperatures ranging from 465° to 485°C which is sufficient to melt lead. Atmospheric pressure was found to be 90 to 92 times greater than that at the earth's surface. Such a pressure would be found on earth at an ocean depth of 900 metres.

Before the advent of **Venera 9** and **Venera 10** many astronomers thought that violent winds of 200 km per hour or more would be continuously sweeping across the surface of Venus and that the thick clouds would keep the surface in semi-darkness. But the Soviet spacecraft reported winds of 3 to 13 km per hour. Such modest breezes indicate that distribution of temperatures over the whole surface of the planet is fairly even, a surprise for those who thought the "day" hemisphere would be much hotter than the "night" hemisphere. Illumination of rocks in the photographs was bright and the rocks were casting sharp shadows.

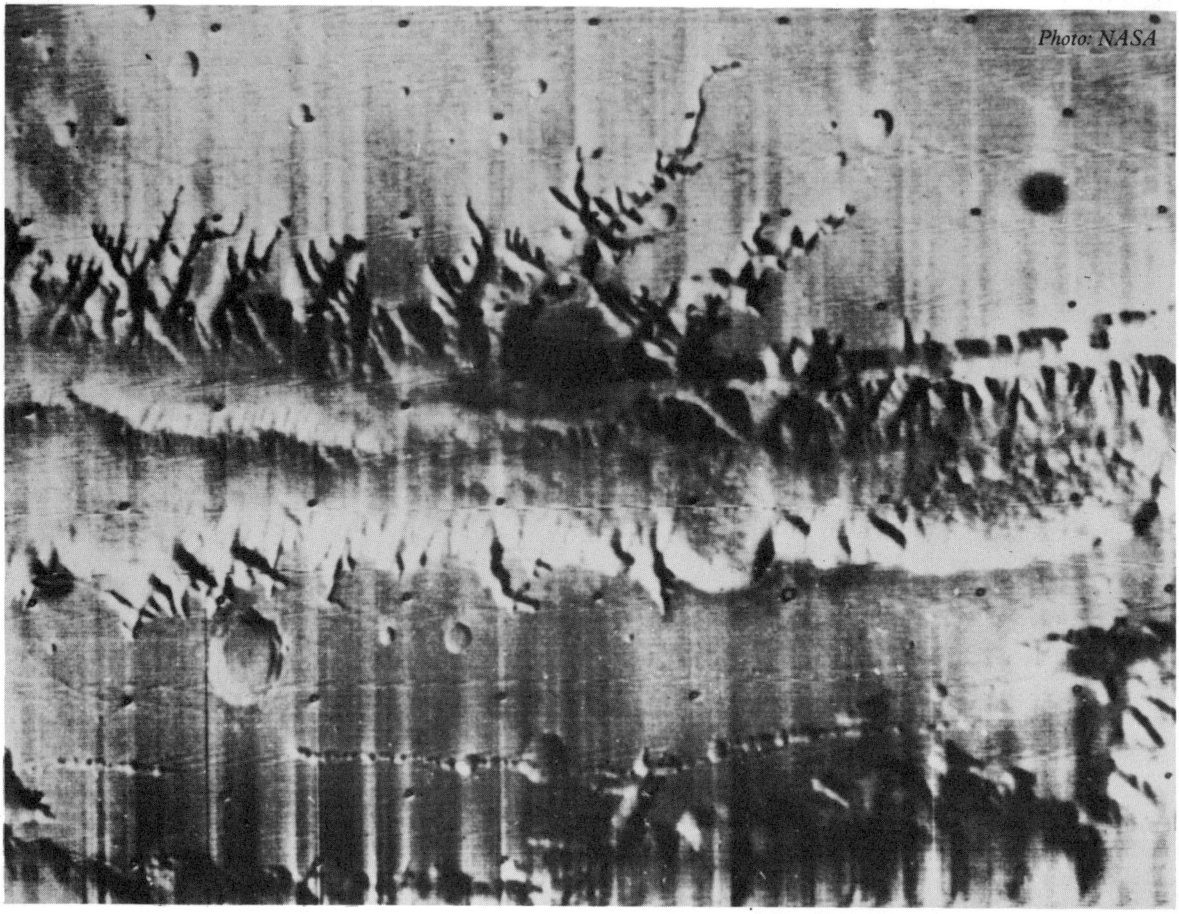

Plate 31. An enormous chasm on the surface of Mars photographed from **Mariner 9** at a height of 2,000 km above the planet. The chasm is 120 km wide and 480 km long. There are branching canyons resembling a network of dry riverbeds but which may be depressions caused by subsidence along lines of crustal weakness.

MARS

Because of its red colour this planet was called Mars after the god of war. For many years astronomers viewing it through telescopes have been tantalised by semi-distinct views of ochre-coloured areas on the planet's surface. Dark patches have been seen, as well as white polar caps which shrink as the Martian winter gives way to spring. Some observers reported what looked like canals but those were probably optical illusions.

Many people, including a few astronomers, dreamed of the day when life might be discovered on Mars, even though the planet was known to be a cold place with only a very thin atmosphere. The possibility was not ruled out that some of the Martian inhabitants might have reached a level of civilisation superior to our own on earth. If Martians were to invade the earth, how would they treat us – with kindness or with enmity?

But in 1965 any hopes of advanced life on Mars were dampened when the spacecraft **Mariner 4** sent back photographs showing the Martian surface as a rocky, cratered wilderness. Later unmanned probes confirmed that discovery.

Then in 1976 world interest was aroused when the American spacecraft **Viking 1** and **Viking 2** put landers on to the Martian surface in a bid to find out whether any form of living organisms at all, even lowly bacteria, are present there or have existed there in the past. Wonderful photographs in colour were obtained and much information received from life-seeking instruments. The result? No life yet detected.

The Martian surface is as forbidding as the surface of the moon, Mercury and Venus but the colour photographs from Mars reveal fine-grained soil largely covered with rocks of different colours – mainly reddish but a few are blue-green and opalescent. The sky shows up with a pinkish hue.

Photographs from the **Viking** orbiters show topographical formations such as deep canyons, dry channels and winding valleys whose general appearance strongly suggest that liquid water once flowed freely on Mars. Even today there is much water locked up in the planet's two polar caps.

The atmosphere consists chiefly of carbon dioxide, about 95 percent, with a little nitrogen and argon. The soil contains phosphates and many other important ingredients. In all probability Mars is capable of supporting life of some sort. Perhaps Mars is not dead, only sleeping.

Mars has a diameter a little more than half the earth's diameter. It has two tiny moons, Phobos, meaning fear, and Deimos, meaning terror. Phobos, photographed by both **Mariner 9** and by **Viking 2,** is a potato-shaped lump of rock about 24 km long and 21 km wide. Its surface has craters up to 5 km wide and there are also mysterious parallel streaks and chains of small craters near its equator.

ASTEROIDS

These are lumps of rocky material ranging in size from dust particles to mini-planets 770 km in diameter. Each one travels in its own particular orbit, the majority keeping within a wide belt between the planets Mars and Jupiter.

The largest and brightest of the asteroids is Ceres discovered by Piazzi in 1801. Next detected were Pallas, Juno and Vesta. The existence of an asteroid is indicated by the appearance of short trails of light in photographs of a star field. At the present time thousands of asteroids are listed.

Some asteroids, for example Apollo (1932), Adonis (1936) and Hermes (1937) have come within 8 million km of the earth. In 1968 Icarus raced past the earth within 6 million km, being spotted by observant astronomers including Albert Jones of Tahunanui, New Zealand. It was also detected by astronomers using radar techniques.

JUPITER

As the giant of the planetary system this planet well deserves the name of Jupiter who in mythology was the lord of heaven and presumed father of gods and men. The planet is large enough to hold over 1000 earths.

A newcomer to astronomy, looking through a telescope at Jupiter, is likely to be amazed that the telescope, even a fairly small one, can transform what appears to be a distant star (778 million km away) into a planet-like object flattened at the poles and exhibiting bands of clouds parallel to the equator. The flattening at the poles is a consequence of Jupiter's very fast spinning motion. The huge object swings round in about 9 hours 55 minutes compared with earth's leisurely rotation of 24 hours. In a fairly large telescope a mysterious Red Spot can be seen on the background of clouds. In a small telescope an observer can see four of Jupiter's 14 moons. Those four were first detected by Galileo

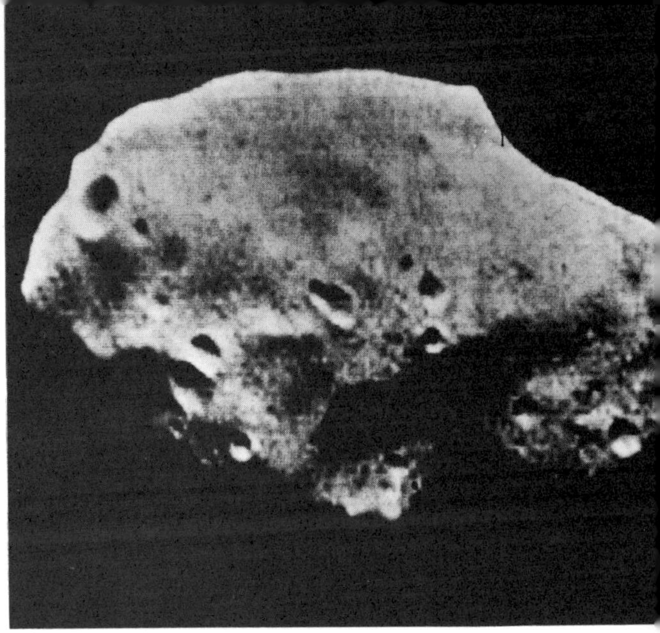

Plate 32. Phobos, a Martian moon, showing indentations probably caused by impacts of fast moving meteoroids. *NASA*

Galilei in 1609. If, reasoned Galileo, a system could exist in which several moons are travelling freely round a central planet, then in a similar system the earth and other planets could be circling the central sun. Galileo was right but his idea met with bitter and prolonged opposition.

When the spacecraft **Pioneer 11** swept past Jupiter in 1975 it relayed back to earth exciting photographs revealing details of the Jovian cloud cover. Study of the photographs indicated that the Great Red Spot is a hurricane-like turbulence contained between bands of cloud flowing in opposite directions.

Jupiter is virtually a gas planet consisting mainly of hydrogen but containing also helium, ammonia and methane. A human being unfortunate enough to be placed a few thousand kilometres below the cloud surface would experience pressures sufficient to crush him beyond recognition.

Surrounding Jupiter is a strong magnetic field. Electrically charged particles streaming through this field posed a hazard to instruments on board the spacecraft **Pioneer 10** and **Pioneer 11.**

The most prominent of Jupiter's 16 moons – Io, Europa, Ganymede and Callisto – are those detected by Galileo and are called the Galilean moons. Two of the Galilean moons are about 1½ times as large as our moon and the other two are almost the same size as ours. The several outermost moons are in retrograde motion and may have been captured by Jupiter at some time after the planet and other moons had been formed.

SATURN

Saturn is a cold, giant world big enough to hold about 700 earths and it is covered with clouds of hydrogen and helium. Its average density is so low that if a water reservoir large enough to hold it could be found, the planet would float with three-tenths of its bulk above water level.

The rings, unique in the solar system, have an overall diameter of 272,000 km and a thickness of only 12 to 24 km. The whole vast, thin, flat system revolves round the planet, not as one solid object but as innumerable separate pieces of material travelling on individual orbits and at different speeds. The pieces may be lumps of rock or perhaps of frozen ammonia.

Saturn has 23 moons. One of these, Titan, is the largest moon in the solar system and the only one known definitely to possess an atmosphere.

URANUS

With its axis inclined at an angle of 98° to the "upright" (the perpendicular to the plane of the ecliptic), Uranus rolls round the sun somewhat like a barrel on its side except that the Uranian axis maintains a constant direction in space. At certain stages during its 84-year journey round the sun one or other of its polar regions points directly to earth. There must be a queer kind of seasons on Uranus. The planet has five moons, one of them nearly as large as ours.

NEPTUNE

At the great distance from the sun of 4496 million km, Neptune is extremely cold and its surface shows few details through the largest telescopes. Its atmosphere contains methane, ammonia and hydrogen.

Each year on Neptune, that is, the time required to circle the sun, lasts 165 earth-years and therefore a human being born there would not live long enough to celebrate his first Neptunian birthday. Neptune has two moons.

PLUTO

Pluto is only a dot in the largest telescopes. Outermost planet in the solar system, and at a mean distance of 5900 million km from the sun, Pluto has been appropriately named after the god of the underworld. The surface must be so cold that all known substances, with the possible exception of helium, are locked in frozen immobility. Its year is 248 earth-years and hence only a small fraction of its journey round the sun has been observed since the planet was discovered in 1930. For that reason the highly elliptical shape of its orbit has not been accurately calculated. Nevertheless, according to one estimate, Pluto will come nearer to the sun than Neptune during the period 1979 to 1999. Pluto has one moon, Charon.

Chapter 13.

HOW TO FIND THE PLANETS

For a number of reasons there is a special fascination about gazing at one of the bright planets.

We can think of it as a member of our own solar system rather than as a more distant and fiercely burning star. There is always the possibility, however slim, that a planet might support some form of life. Unmanned spacecraft have landed upon or flown near some of the planets. Man himself might some day set foot on a planet, for example Mars.

Through a telescope of suitable size we can obtain an enlarged view of the planets, in some cases being able to observe details of their surfaces and to detect some moons in orbit round the parent planet.

Even with the naked eye we can follow the progress from night to night of the five brightest planets as they appear to move slowly against the background of stars as if asserting their independence among myriads of other celestial objects.

PATHS OF THE PLANETS

All the planets, as well as the sun and the moon, move across the sky within an imaginary band known as the **zodiac**. This band extends about 8° (roughly the width of your fist held at arm's length) on each side of the **ecliptic,** the apparent path of the sun. Fig. 26 page 37 shows the ecliptic on the celestial sphere. When, in imagination, the celestial sphere has been sliced into three sections (two polar and one equatorial), flattened to form the Whole-Year Charts I to IV, the ecliptic shows up as the curved line which is labelled on Whole-Year Charts II, III and IV showing the equatorial regions. Here the boundaries of the zodiac can be envisaged as two lines drawn parallel to the ecliptic, one line 8° above it, the other 8° below it. The width of the zodiac can be estimated by consulting the vertical scale of degrees at each end of these three charts.

Suppose you saw a bright starlike object very close to the star Aldebaran and just below it (see Aldebaran in Whole-Year Chart II). Because the unknown bright object is not in the chart you suspect it to be a planet. Comparing the chart with the sky, you would note that the object lies within 8° of the ecliptic, that is, it is in the zodiac. If it had not been within the zodiac or very close to it, the object could not have been a planet.

From the descriptions of the individual planets as given later in this chapter you could almost certainly identify the object near Aldebaran. Of course, a few nights later you would discover that the object had changed its position though keeping within the zodiac.

CONSTELLATIONS OF THE ZODIAC

The zodiac contains the following constellations. For readers wishing to commit them to memory in the correct order, a mnemonic is supplied.

1. **Aries**	A	7. **Libra**	likely
2. **Taurus**	tall	8. **Scorpius**	skimmed
3. **Gemini**	gentleman	9. **Sagittarius**	some
4. **Cancer**	called	10. **Capricornus**	cream
5. **Leo**	Leo	11. **Aquarius**	and
6. **Virgo**	very	12. **Pisces**	partook

Part of another constellation, Ophiuchus, extends into the zodiac. For the English equivalents of these names, refer to Appendix D. The boundaries of these constellations do not correspond to those of the so-called Signs of the Zodiac which were introduced about 2100 years ago.

TO FIND A PARTICULAR PLANET

Some newspapers, observatories and astronomical societies publish at regular intervals information enabling anyone to find individual planets in the sky. Such information usually includes mention of where and when to look for particular planets. It may give the approximate times when each planet rises (approximately east) and sets (approximately west). The naked eye appearance of the planets is usually stated. Enquire at your nearest public observatory for newsletters, bulletins or handbooks dealing with the month by month position of the planets.

GENERAL ADVICE FOR LOCATING PLANETS

Mercury: Because this planet's orbit is closer to the sun than that of the Earth, we have to look somewhere near the sun to find it. Thus Mercury can most easily be located just before dawn in the eastern sky or just after sunset in the western sky. At such times it is never higher than 28° above the horizon. It is usually rather faint but sometimes it is brighter than the star Sirius. Astronomical bulletins occasionally indicate its position relative to some other easily identifiable object such as the moon or Venus.

Venus: Venus is the second planet outwards from the sun and, like Mercury, it is best seen before dawn in the eastern sky or after sunset in the western sky. It can appear higher above the horizon than can Mercury and, at its brightest, it outshines all other planets.

Through a telescope both Mercury and Venus show phases resembling the phases of the moon, that is, crescent, first quarter and so on. Fig. 31 shows some observational aspects of Venus. Such aspects apply also to Mercury.

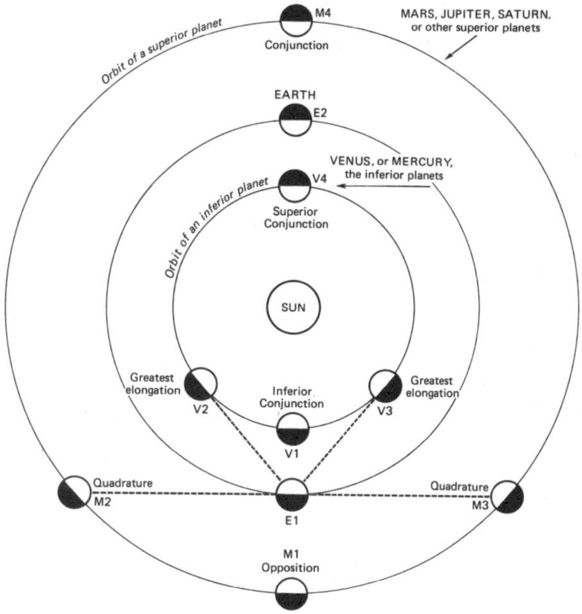

Fig. 31. Observational aspects of the planets. An observer on the earth at E1, looking at Venus, can never see it further from the sun, in the line of sight, than at its positions of greatest elongation. A similar restriction applies to Mercury, the other inner planet. At inferior or superior conjunction, Venus and Mercury are too close to the sun, in the line of sight, to be observable. For an observer at E1, Mars is favourably situated when at opposition, M1, and this is also true for Jupiter and the other planets.

Mars: The red colour of Mars is a valuable aid to its recognition. When at opposition (see Fig. 31) it can be seen at its brightest and this is the time when astronomers study the planet under the most favourable conditions. Hazy patches on the Martian surface, as well as white polar caps, can be detected through a telescope of 75 mm aperture, but for serious observations a larger instrument is required. Sometimes temporary markings which we now know are dust storms can be recognised.

Asteroids: Special charts are needed to pinpoint some of the largest asteroids. A series of photographs will show that any particular asteroid is changing its position relative to neighbouring stars.

Jupiter: This is a favourite object for observation. Its slow progress across the starry background can be traced for weeks at a time with the naked eye. In a 60 mm telescope greyish zones parallel to the equator are visible and the flattening of the polar regions is readily noticeable.

The four pinpoints of light near Jupiter are its brightest moons — Io, Europa, Ganymede and Callisto. These can also be seen in binoculars which must for this purpose be held steady by resting them on a support such as a fence post. Movements of the moons become apparent if observed at intervals of a few hours. Keen observers can sometimes witness such fascinating happenings as an **occultation** (disappearance of a moon behind the planet's disk), a **transit** (a moon passing in front of the planet, sometimes casting a shadow on the planet) or, rarely, the **eclipse** of one moon by another moon. Predictions of these phenomena are published in the monthly issues of the magazine *Sky and Telescope.*

Saturn: Most people seeing Saturn in a telescope for the first time are thrilled by the experience. The planet and its ring system give the impression of a white glowing jewel displayed on dark velvet (the black sky). A 60 mm telescope can provide this view and it will show Saturn's brightest moon, Titan.

In a larger telescope, say a 100 mm refractor, more details are revealed. These include Cassini's Division, a dark line separating two of the three main rings of the ring system.

As Saturn makes its relatively slow journey round the sun its ring system presents itself in different attitudes. First we see the south side of the rings, then the edge of the rings when they become almost invisible, then the north side. One whole cycle of changes takes 29½ years.

Uranus: Because of its great distance from us, nearly 3000 million km, we need a special chart to locate Uranus. When found, it can just be detected with the naked eye in good seeing conditions but binoculars bring it more comfortably into view. In a large telescope it shows a pale greenish colour with very little detail on the surface. Its five moons are very faint.

Plate 33. Comet Bennett photographed on 21 March 1970. Note the straight gas tail and the curved dust tail. *B. Ward*

Neptune: An amateur with a small telescope can find Neptune, using a special chart. It appears as a faint starlike object. The two moons are not visible except in large telescopes.

Pluto: Comparatively few amateur astronomers have seen Pluto. Of course a large telescope and special chart are required. To convince himself that he had really detected the planet an observer would probably photograph the star field containing the object and then take another photograph some months later to ensure that the object had moved against the background of stars at the rate appropriate for such a remote planet.

Chapter 14.

COMETS AND METEORS

COMETS

A typical comet seen by the naked eye is a somewhat awesome object consisting of a fuzzy bright head and a very long fainter tail. The head contains a nucleus made probably of cohesive rocky material and frozen gases. The tail, so tenuous that stars can be seen on the far side of it, consists of gases and dust. A tail is sometimes many millions of kilometres long.

Some comets are on orbits which take them round and round the sun for many years. Others make one circuit of the sun and then disappear, possibly leaving the solar system for ever.

The most famous of all comets, Halley's Comet, has made many appearances and is expected to return in 1986.

METEORS

Innumerable pieces of rock, ranging from huge chunks (weighing many tonnes) down to grain-sized particles, are travelling at great speed through the solar system. They are called **meteoroids.** Those that strike our atmosphere are raised to white heat by friction and become visible as **meteors,** sometimes wrongly referred to as shooting stars. At this stage they may soon become extinguished and slowly float as ashes to the ground. Alternatively, if large enough to survive their journey through the atmosphere, they drop on to the ground and are then known as **meteorites.**

Australia's heaviest recorded meteorite, of 12 tonnes, was found in 1966 near Forrest, in Western Australia. It is now in the Western Australian Museum. New Zealand's biggest meteorite was discovered near Dunganville not far from Greymouth in 1976. Consisting chiefly of iron, it weighed 54 kg.

VIII. BEYOND THE SOLAR SYSTEM

Chapter 15.

ORDINARY STARS, WHITE DWARFS, NEUTRON STARS (PULSARS); BLACK HOLES; QUASARS; COSMOLOGY

Modern astronomy is no longer restricted to just what we can see with the unaided eye and to what can be "brought closer to us" by means of an optical telescope. In other words we do not need to rely only on light waves and the messages they bring from distant parts of the universe.

There are other types of electromagnetic waves and for each type a new branch of astronomy has been developed. Thus, taking each group of waves in turn from those with the shortest wavelength to those with the longest wavelength, we have gamma ray astronomy, X-ray astronomy, ultraviolet ray astronomy, light ray or optical astronomy, infrared ray or heat wave astronomy, and radioastronomy.

There is another branch, neutrino astronomy, which may be mentioned in a class of its own because we are not quite sure whether neutrinos should be regarded as waves or particles. For description of neutrinos, refer to Chapter 9 page 45.

Plate 34. The 64 m radiotelescope at Parkes, Australia. It established communication with the **Apollo 15** astronauts on the moon, which can also be seen in this picture. *CSIRO*

Thus it will be realised that astronomical knowledge is advancing on a very broad front, probing some of Nature's deepest mysteries.

An ordinary star like most of the stars we can see in the sky, survives in a stable form during the period in which its chief fuel, hydrogen, is being transformed into helium. But after a few thousand million years of "life" the star runs short of hydrogen. The remaining nuclear reactions are no longer able to provide enough thrust to resist gravitational forces acting inwards. The star therefore contracts, thereby generating enough heat to ignite the remaining hydrogen shell surrounding the core. The star now expands greatly and cools. It becomes a red giant. In due course all its nuclear fuel is exhausted and it then contracts drastically to become a **white dwarf,** resembling Sirius B as described in Chapter 4, page 14. A white dwarf has no future except to cool into a black dwarf, which, if it does not collide with another sizeable object, will remain invisible forever.

A neutron star or pulsar is the superdense remnant of a star which, having a mass about twice that of an ordinary star and having reached the giant stage in its evolution, collapses and then explodes catastrophically as a **supernova.** In the explosion the supernova flings about 90 percent of its substance into surrounding space while outshining for a few days even an entire galaxy of stars.

The central remnant of the explosion is an object composed of neutrons which are sub-atomic particles. They are compressed to a density about 100 million times greater than that of a white dwarf.

In 1967 a mysterious object was discovered by A. Hewish and his colleagues at Cambridge University, England. It was emitting pulses of radio waves with an extraordinary degree of regularity and, like many other similar objects discovered later, it has become known as a **pulsar.** In all probability a pulsar is a neutron star about 10 km in diameter, rotating very rapidly and "beaming" pulses of radio waves, just as a lighthouse swings a beam of light round and round.

A black hole is a small region, only a few kilometers in diameter, into which a star more than three times as massive as the sun has collapsed after a supernova explosion. The material within a black hole is dense almost beyond belief. The "surface" of a black hole is called the absolute-event horizon.

The gravitational field within the absolute-event horizon is so strong that no material particle or electromagnetic radiation such as light could escape from it. Thus a black hole is invisible and the only evidence of its existence is the gravitational influence it exerts on neighbouring material. Astronomers are not all agreed that black holes actually exist.

A quasar is a small starlike object with such an extraordinary set of properties that its exact nature is not fully understood. Light received from it indicates that it is extremely remote – at the very boundaries of the known universe. Yet it is emitting radio waves of greater intensity than can be explained. It is giving off as much light as does a whole galaxy of stars.

One possible though incomplete explanation is that a quasar is a super-violent explosion at the centre of certain types of galaxies classified as N galaxies and Seyfert galaxies.

A galactic nebula is a vast cloud-like formation of gas and dust located within the arms of the Milky Way Galaxy. An example is the Great Nebula of Orion.

Cosmology is the study of the universe considered as a whole. It deals with the life history of galaxies and such questions as the origin, present structure and future development of the universe itself.

Plate 35. The Andromeda galaxy and its two satellite galaxies.

Hale Observa

APPENDICES

APPENDIX A

CHOOSING BINOCULARS

For general observation of either land or sky objects a pair of binoculars labelled 7 × 50 can be recommended. The 7 × means a magnification of 7 times the diameter of an observed object, and the 50 indicates that the aperture (effective diameter) of each main lens at the wide end of the instrument is 50 mm. A labelled field of 7.1° is very satisfactory and it means that the eye will see through the binoculars a circular field of sky 7.1° in diameter.

A higher magnification than 7 × can be useful for the observer doing an extended programme of research but he might find it necessary to support the binoculars on a tripod to hold them steady.

I often use, for tourist sightseeing, as well as casual observations of sky objects, a very small pair of binoculars specified as 10 × 20, field 4.2°. Having overall dimensions of a little less than 9 cm by 9 cm they fit conveniently into a shirt pocket and are ready for use within seconds.

CHOOSING A TELESCOPE

A new telescope capable of giving a satisfactory performance is quite a costly item ranging from about $200 to more than $1000. Consequently I recommend newcomers to defer purchasing a telescope until they have done some viewing through someone else's. They can then decide with confidence whether they are likely to enjoy owning one for themselves. Private owners of telescopes can be contacted through your nearest astronomical society.

Occasionally a second-hand telescope in good order can be purchased at a much reduced price. Such an instrument, however, should be tested for its performance by a well-qualified astronomer.

Telescopes are of two main types –

Refractors: They gather incoming light with a **lens.** (Plates 13 and 36).

Reflectors: They gather incoming light with a **concave mirror.** (Plates 37 and 39).

Each type has advantages and disadvantages. A refractor requires very little maintenance but a

Plate 36. A 60 mm refractor with altazimuth mounting. *A. L. Rounthwaite*

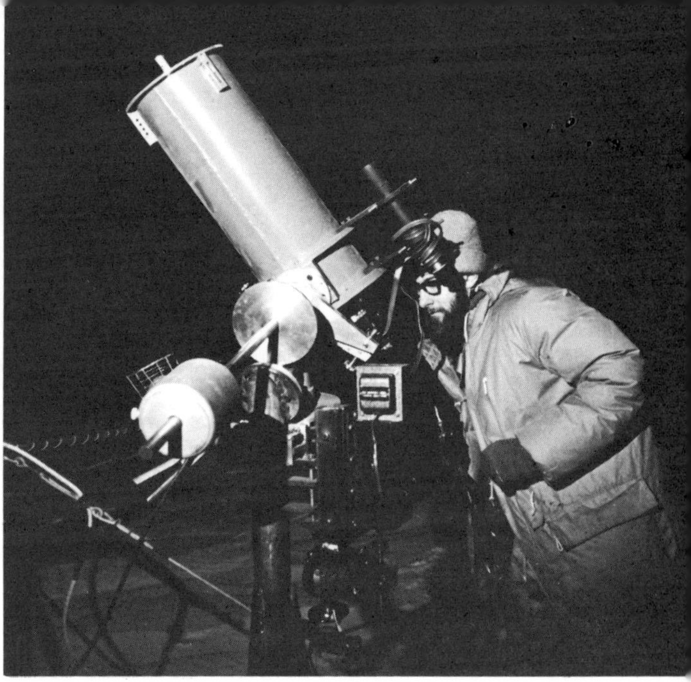

Plate 37. New Zealand scientist John Elder, at Scott Base in the Antarctic, testing the feasibility of making astronomical observations in conditions of extreme cold. *R. M. McBride*

reflector must have its mirror re-aluminised every few years. A modified form of reflector using the Cassegrain optical system is a more compact instrument than the simpler Newtonian form. Some Cassegrain reflectors have their mirrors completely sealed from contact with the outside atmosphere and, though costing about $1000 or more, they are very fine telescopes indeed.

Small telescopes of about 60 mm aperture usually have an **altazimuth mounting.** This permits an up-and-down movement and a sideways movement of the tube, both of which must be constantly activated by hand to follow a star or planet as it moves along a circular path in the heavens.

Larger telescopes usually have an **equatorial mounting** which has the advantage that, after the telescope has been properly set up, only one motion is required, the motion obtained by rotating the tube round its polar axis. That motion can be sustained with the hand or with an automatic clock drive.

I recommend a 60 mm refractor for beginners. Any telescope aperture less than 60 mm is likely to prove a disappointment in spite of any claims as to its magnifying power.

Another way of acquiring a telescope is to make one for yourself. With practical guidance from a knowledgeable member of your nearest astronomical society you could construct a reflector telescope of aperture 150 mm or higher, at a small fraction of the cost of a professionally made refractor.

Plate 38. Setting-circles on equatorial mounting of a 102 mm refractor. *A. L. Rounthwaite*

Plate 39. Mr G. L. Loftus, using a 50 cm paraboloidal mirror which he had ground and polished, demonstrates that the mirror can focus the feeble heat of a winter's sun to ignite an oily rag. Illuminated particles of smoke show that the sun's light is also being focused.

APPENDIX B

ASTRONOMICAL SEASONS

Commencing dates:

Summer	22 December Sun over Tropic of Capricorn Solstice
Autumn	21 or 22 March Sun over Equator Equinox
Winter	22 June Sun over Tropic of Cancer Solstice
Spring	22 or 23 September Sun over equator Equinox

APPENDIX C

ASTRONOMICAL UNITS OF DISTANCE, SPEED OF LIGHT

Light-year: the distance that light travels in one year. It is equal to 9.4607×10^{12} km or 9.4607 million million km.

Astronomical Unit: the mean distance between the earth and the sun. It is 1.496×10^{8} km or 149.6 million km.

Parsec: the distance at which the radius of the earth's orbit subtends an angle of 1 second of arc. The word parsec is a contraction of parallax and second. It is 30.857×10^{12}km = 206,265 a.u. = 3.2616 l.y. The term parallax here means the small angle subtended by earth's orbit. Refer fig. 65.

Speed of light in a vacuum: 299,792.5 km per second.

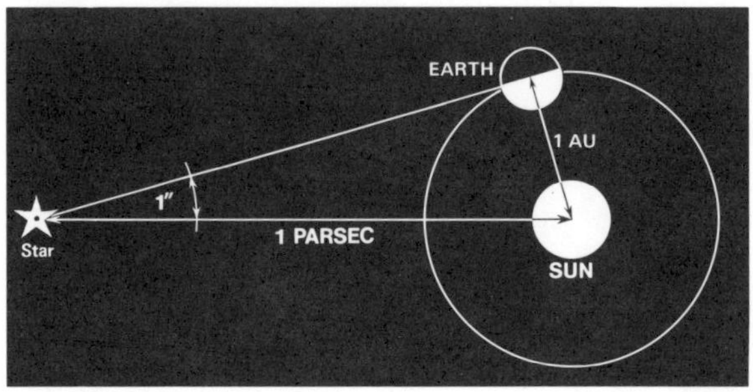

Fig. 32. The parsec is the distance at which the mean radius of the earth's orbit (called the Astronomical Unit or AU) subtends an angle of 1 second of arc. The parsec = 3.2616 light-years.

Fig. 33. The seasons. Because of the tilt of the Earth's axis, the spinning Earth presents different attitudes to the sun as it travels round the sun each year. This situation gives rise to the seasons: summer, autumn, winter and spring. In the northern hemisphere the seasons are the reverse of those in the southern hemisphere.

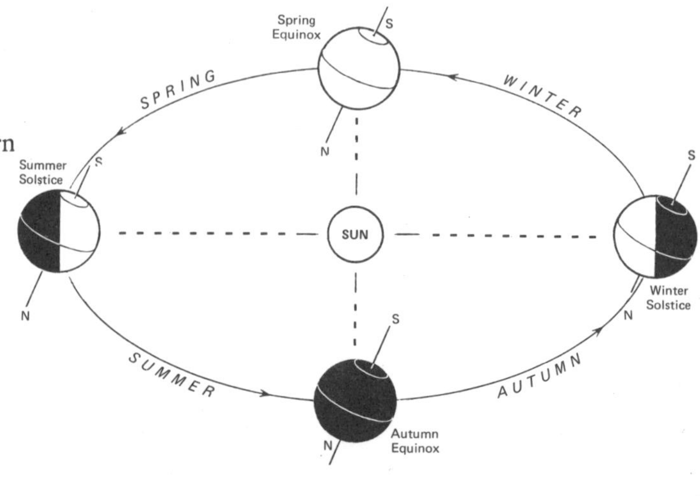

APPENDIX D

SOME CONSTELLATIONS

Abbreviation	Latin Name	Possessive Case	English Equivalent
And	Andromeda	Andromedae	Andromeda
Aqr	Aquarius	Aquarii	Water Carrier
Aql	Aquila	Aquilae	Eagle
Aur	Auriga	Aurigae	Charioteer
Boo	Boötes	Boötis	Herdsman
Cnc	Cancer	Cancri	Crab
CMa	Canis Major	Canis Majoris	Larger Dog
CMi	Canis Minor	Canis Minoris	Smaller Dog
Cap	Capricornus	Capricorni	Goat
Car	Carina	Carinae	Keel
Cen	Centaurus	Centauri	Centaur
Cet	Cetus	Ceti	Whale
Com	Coma Berenices	Comae Berenices	Hair of Berenice
CrA	Corona Australis	Coronae Australis	Southern Crown
CrB	Corona Borealis	Coronae Borealis	Northern Crown
Crv	Corvus	Corvi	Crow
Cru	Crux	Crucis	Cross
Cyg	Cygnus	Cygni	Swan
Eri	Eridanus	Eridani	River
Gem	Gemini	Geminorum	Twins
Gru	Grus	Gruis	Crane
Leo	Leo	Leonis	Lion
Lep	Lepus	Leporis	Hare
Lyr	Lyra	Lyrae	Lyre
Mus	Musca	Muscae	Fly
Oct	Octans	Octantis	Octant
Oph	Ophiuchus	Ophiuchi	Serpent Holder
Ori	Orion	Orionis	Orion
Pav	Pavo	Pavonis	Peacock
Peg	Pegasus	Pegasi	Pegasus
PsA	Piscis Austrinus	Piscis Austrini	Southern Fish
Pup	Puppis	Puppis	Stern
Sgr	Sagittarius	Sagittarii	Archer
Sco	Scorpius	Scorpii	Scorpion
Tau	Taurus	Tauri	Bull
Tuc	Tucana	Tucanae	Toucan
Vel	Vela	Velorum	Sails
Vir	Virgo	Virginis	Virgin

APPENDIX E

THE GREEK ALPHABET AS USED IN NAMES OF STARS

Alpha	α	Epsilon	ε	Iota	ι	Nu	ν	Rho	ρ	Phi	φ
Beta	β	Zeta	ζ	Kappa	κ	Xi	ξ	Sigma	σ	Chi	χ
Gamma	γ	Eta	η	Lambda	λ	Omicron	ο	Tau	τ	Psi	ψ
Delta	δ	Theta	θ	Mu	μ	Pi	π	Upsilon	υ	Omega	ω

With a few exceptions such as Pollux (Beta Geminorum) and Castor (Alpha Geminorum), the brightest star in a constellation is designated alpha, the next brightest beta, and so on in the order given in the list above.

APPENDIX F

ELEMENTS OF THE PLANETARY ORBITS

(From *Handbook of the British Astronomical Society*)

Name	Equatorial diameter in km	Density (water = 1)	Escape Velocity in km sec^{-1}	Volume (Earth = 1)	Surface Gravity (Earth = 1)	Mean Distance in million km	Sidereal Period (length of "year" in days)
Sun	1,392,000	1.409	617.5	1303.600	27.90		
Moon	3,476	3.342	2.38	0.0203	0.1653		
Mercury	4,880	5.5	4.3	0.056	0.381	57.91	87.969
Venus	12,100	5.25	10.36	0.8572	0.9032	108.21	224.701
Earth	12,756	5.517	11.18	1.0000	1.0000	149.60	365.256
Mars	6,790	3.94	5.03	0.1504	0.3799	227.94	686.980
Jupiter	142,800	1.330	60.22	1318.7	2.643	778.34	4,332.59
Saturn	119,300	0.706	36.26	743.6	1.159	1,427.01	10,759.20
Uranus	47,100	1.71	22.5	47.1	1.11	2,869.6	30,684.9
Neptune	48,400	1.77	23.9	53.7	1.21	4,496.7	60,190.2
Pluto	5,900	4.5	4.2	0.10	0.47		

INDEX